ADELPHI

Paper • 300

The Politics of Oil in the Caucasus and Central Asia

Contents

Oxford University Press, Walton Street, Oxford OX2 6DP
Oxford New York
Athens Auckland Bangkok Bombay
Calcutta Cape Town Dar es Salaam Delhi
Florence Hong Kong Istanbul Karachi
Kuala Lumpur Madras Madrid Melbourne
Mexico City Nairobi Paris Singapore
Taipei Tokyo Toronto
and associated companies in
Berlin Ibadan

Oxford is a trade mark of Oxford University Press

Published in the United States
by Oxford University Press Inc., New York

© The International Institute for Strategic Studies 1996

First published May 1996 by Oxford University Press for
The International Institute for Strategic Studies
23 Tavistock Street, London WC2E 7NQ

Director: Dr John Chipman
Deputy Director: Rose Gottemoeller

British Library Cataloguing in Publication Data

Data available

Library of Congress Cataloging in Publication Data

ISBN 0-19-828092-0
ISSN 0567-932X

GLOSSARY

AIOC	Azerbaijan International Operating Company
b/d	barrels per day
CC	Caspishelf Consortium
CCO	Caspian Cooperation Organisation
CIS	Commonwealth of Independent States
CPC	Caspian Pipeline Consortium
CSCE	Conference for Security and Cooperation in Europe
EBRD	European Bank for Reconstruction and Development
ECO	Economic Cooperation Organisation
EU	European Union
GIOC	Georgian International Oil Corporation
IFC	International Finance Corporation
IFI	International Financial Institutions
IMF	International Monetary Fund
NGO	Non-Governmental Organisation
OSCE	Organisation for Security and Cooperation in Europe
SOCAR	State Oil Company of the Azerbaijan Republic
TPAO	Turkish Petroleum Company
UN	United Nations
USSR	Union of Soviet Socialist Republics

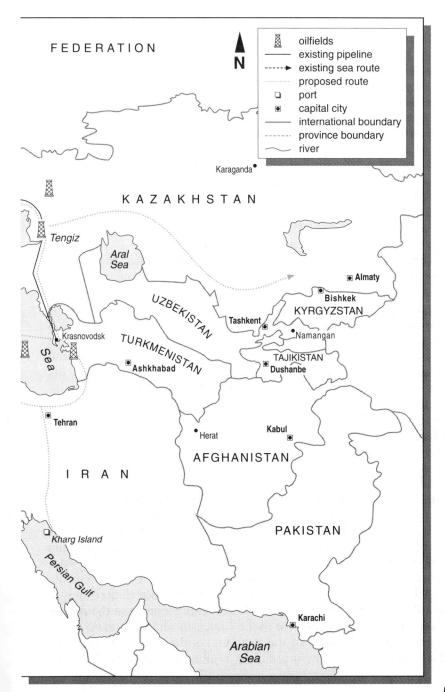

FEDERATION

N

🗼	oilfields
——	existing pipeline
----►	existing sea route
....	proposed route
❑	port
◉	capital city
——	international boundary
·········	province boundary
～～	river

Karaganda ●

K A Z A K H S T A N

🗼 Tengiz

Aral Sea

◉ Almaty

◉ Bishkek

KYRGYZSTAN

UZBEKISTAN

Tashkent ◉

● Namangan

● Krasnovodsk

TURKMENISTAN

TAJIKISTAN

◉ Ashkhabad

Dushanbe

🗼 Sea 🗼

◉ Tehran

● Herat

Kabul ◉

AFGHANISTAN

I R A N

PAKISTAN

❑ Kharg Island

Persian Gulf

Karachi ◉

Arabian Sea

INTRODUCTION

With billions of dollars and crucial strategic influence at stake, the struggle for control over the vast oil resources in the Caucasus and Central Asia is a tale of political intrigue, fierce commercial competition, geo-strategic rivalries, ethnic feuding and elusive independence. Energy resources in this region are concentrated mainly in the Caspian Sea Basin, in Kazakhstan, Azerbaijan and Turkmenistan, but some of the area's other states, such as Georgia and Armenia, also play a role in energy export issues. Straddling Europe, the Near East and Asia, the Caspian region is one of the largest unexploited sources of oil in the world. Proven and possible reserves are estimated to be as high as 200 billion barrels, putting the region on a par with Iraq.[1] In addition, the area is rich in natural gas with estimated proven and possible reserves of up to 7.89 trillion cubic metres – as much as those of the US and Mexico combined.[2]

The demise of the Soviet Union in 1991 and the subsequent rise of the more vulnerable and less experienced newly independent states of the Caucasus and Central Asia led to an intense political and commercial competition for control over oil resources and export. Some analysts have compared this situation to the 'Great Game' – a nineteenth-century rivalry between Victorian England and Tsarist Russia. The matrix of national identities, mentalities, goals and instruments, however, has changed significantly. In addition, the new players differ in their perception of the game, with some maintaining that the competition is no longer a zero-sum game, while others still believe that it is and see the world through a traditional balance-of-power framework. The stakes involved, however, remain unchanged – power, influence, security, wealth.

The new playing field is inherently complex and is further complicated by a vast array of problems. Within the region, these include intra-regional conflict, internal political instability, unscrupulous entrepreneurial operators, and a shortfall in commercial expertise and legal infrastructures. Beyond the region lie the threats of proprietorial and competing neighbours. Kazakhstan, Azerbaijan and Turkmenistan, which share the majority of the region's energy deposits, are landlocked and, therefore, dependent on their immediate neighbours for export. This makes them vulnerable to their neighbours' problems and, sometimes, to becoming a pawn in the rivalry of the larger powers adjacent to the Caucasus and Central Asia.

The mercurial nature of the structural and political evolution of these states and of the surrounding regions makes formulation of long-term regional objectives by third-party policy-makers and Western businesses extremely difficult. Both stand to gain or lose significantly depending on whether they correctly predict the outcome of the region's many developing issues.

The benefits of developing and exploiting the Caspian region's oil resources are clear. First, the margin between world oil production capacity and world demand is projected to narrow in the next decade, leading to greater dependence on the Persian Gulf.[3] Central Asian oil could offer an important alternative, diversifying supply. In consequence, as a powerful geo-strategic key, oil offers the region's states the wherewithal to exploit their best opportunity for true independence in 70 years. Finally, with a possible 90–200bn barrels, the potential for national and commercial profit is substantial.[4]

In a wider context, Caspian oil is tied to, and will affect, issues central to current and developing international relations. These include:

- the political and economic future of Russia, and its behaviour towards neighbours and former Soviet republics;
- the political and economic future of Turkey;
- Iran's position in the region, and its relations with the West, with Russia, and with its other neighbours in the former Soviet Union;
- the strategic consequences of greater dependence on Persian Gulf oil;
- tension between Pakistan and India;
- China's future policy towards its neighbours;
- the potential spread of Islam to the region.

This paper focuses on the Caucasus and Central Asian region as an oil producer of considerable geo-strategic importance. Looking first at the region's history and the strategic role that oil has played since the nineteenth century, the paper goes on to identify the major locations of the region's oil and the factors affecting oil development, exploitation and export. These factors include: the political interests and policies of external parties (Russia, Turkey, Iran, China, Pakistan and the US, along with other developed countries); the Caspian Sea dispute – which, in a sense, encapsulates the nature of the intra-regional competition; the internal political problems of the region's states; and the technical and commercial impediments to implementing oil projects. After covering the largest oil development projects as specific cases, the paper discusses principal short-

and long-term export options, including the complex issue of pipelines, and concludes by suggesting how the regional states, the US, Russia and other key actors could develop their policies to encourage stable exploitation of Caspian oil.

I. THE HISTORY OF OIL IN THE REGION

The presence of oil in the Caucasus and Central Asia is recorded as far back as the thirteenth century. Throughout the twentieth century, Caspian oil has played a key strategic role in world politics, frequently the source of contention between external superpowers. The nineteenth-century Great Game had been based on competition for wider power and influence by asserting control over the Central Asian region. But by the end of the nineteenth century, with technology increasingly capable of exploiting the reserves, oil emerged as a pivotal factor in the competition, and the Game intensified. The Caucasus and Central Asia was no longer just a point of access to the riches of South Asia, in particular India, but a lucrative prize in itself.[1] In fact, the mineral wealth of Central Asia in Turkmenistan and Kazakhstan was not really discovered or exploited on a large scale until the 1950s. From the mid-nineteenth to the mid-twentieth century, most of the competition for oil took place over reserves in the Caucasus region of the Caspian.

Thus, Caspian oil acquired the role of key strategic asset, playing an important part in determining the shape of the modern political landscape. In the late 1800s, great oil barons of the day – the Nobel brothers, the Rothschilds, and the leaders of Royal Dutch Shell – helped Russia to develop Caspian oil resources. The Nobel Brothers' Petroleum Production Company was considered 'one of the greatest triumphs of business enterprise in the nineteenth century'.[2] Indeed, the exigencies of shipping oil out of the region obliged Ludwig Nobel, in 1878, to invent the first operational oil tanker.[3] During that period, Caucasian oil made up 30% of the world oil trade.

This oil carried considerable strategic weight in both World Wars. During the First World War, the Germans, having exhausted their own fuel supplies, tried to seize oil in the Baku region to feed the continuing war effort. In summer 1918, the *New York Times* argued that safeguarding Caucasian oil-fields should be a priority for the Allies, and that they had to be prepared to devote significant military force to this project. As the German military machine ran down, Baku instead fell under Turkish, and eventually Soviet, influence. Denied Baku's precious oil, the Germans were unable to continue the war and surrendered in November 1918.

In the Second World War, Hitler seems to have been determined to use Caucasian oil to fuel his military expansion. After the conclusion of the 1939 Nazi–Soviet Pact, Soviet oil from the Caucasus provided no less than one-third of Germany's imports. In 1941,

Germany launched a series of campaigns to take outright posses-
sion of the region and its mineral reserves. These campaigns
reached their height in 1942 when Hitler stressed at a staff meeting
of the Army Group South that if he failed to gain control of the oil
in the Caucasus, he would be forced to end the war.[4] But the
German campaigns failed for several reasons, including the mountainous terrain, the strength of Soviet defences, and the exhaustion
of German forces, dispersed far from their fuel and food supplies.
As Daniel Yergin points out, 'the Germans ran out of oil in their
quest for oil'.[5]

Great-Game thinking, from the late nineteenth century to the
Second World War, defined oil as a strategic raw material to be
monopolised. During this period Russia, Turkey and the West were
engaged in intense competition for influence over the oil-producing
areas. Internally, the Caucasus and Central Asia was fraught with
tensions and bloody confrontations between the Turkic and Armenian ethnic groups, widespread corruption and poor administration,
cut-throat competition between large oil companies, and the prolif-
eration of dubious international entrepreneurs. The danger of historical parallels notwithstanding, there are clear similarities between
then and now – particularly in commercial competition, corruption,
poor administration and ethnic tension.[6]

In terms of commercial competitiveness, as far back as 1895,
Russia, fearing overwhelming Western – and particularly US –
control over its oil markets, deliberately undermined a substantial
deal in the region between the domestic oil company, American
Standard Oil, the Rothschilds and the Nobels.[7] Another similarity
with present conditions was the frequent Western complaint in the
1890s of arbitrary Russian transport rates and regulations, and
widespread government corruption that made it very difficult to
conduct stable business ventures in the region.[8]

Present inter-ethnic conflict, especially hostility between Turkic
and Armenian populations, derives from a centuries-old tradition. In
1905, trying to divert the blame for harsh working and living conditions, the Russian government fuelled ethnic discord between Tatars
and Armenians, resulting in a massacre of Armenians by Tatars
while Russian troops looked on. Subsequently, however, the Tatars
turned on the Russian government and its oil industry. On the other
side, after their betrayal by the Russians, the Armenians threw their
weight behind Lenin and his followers.[9] In 1918, when the British
withdrew from Baku, the Turks took the area, supporting a local
Muslim faction which, still harbouring resentment from the 1905

period, attacked and massacred the Armenians again.[10] These events provide the backdrop for the conflict that has erupted between Azerbaijan and Armenia over Nagorno-Karabakh more than 80 years later.

For centuries, oil has been the focal point for regional and international competition in the Caucasus and Central Asia. The media of competition, both from the point of view of external and internal players, have been established over years of hot and cold conflict. Natural competition in the area continues to be inflamed by interethnic fighting, administrative corruption and underdeveloped commercial and legal practices. These will undoubtedly affect future ventures in the strategic and commercial development of Caspian oil.

Major Oil Deposits
The largest oil deposits in the Caucasus and Central Asia are located in Kazakhstan and Azerbaijan. Other deposits and smaller projects exist throughout the region, in Georgia, Uzbekistan, Turkmenistan and Armenia. Estimates of proven and possible reserves across the entire area run to 200bn barrels of oil product. This includes about 30bn barrels of discovered reserves, approximately equal to those of the North Sea.[11] Post-Soviet regional accessibility has led to substantially increased involvement on the part of Western businesses. The more sophisticated technology and geophysical expertise they brought with them have produced reserve estimates much higher than official Soviet figures, which were based on explorations conducted 40 years ago, before the development of advanced methods for finding and exploiting deeper deposits. The region's strategic importance increases with the growth in its discovered oil.

The Caspian Sea area is particularly rich in oil deposits. Recent geophysical estimates indicate that the area holds far more than the Soviet estimate of 10bn barrels.[12] The sea's southern end has attracted most exploration because, until recently, the countries governing the region did not have access to the technology necessary to analyse the north, which suffers from poor infrastructure and bad weather. Additionally, the hydrocarbon reservoirs are very deep, and strong currents make geophysical work difficult.

Kazakhstan has much larger reserves than were estimated during the Soviet period. The 12 sedimentary basins in eastern and central Kazakhstan 'possess the characteristics of world-class basins and some have the potential to contain giant oil- and gas fields', in

11

addition to those already discovered.[13] After Russia, Kazakhstan is considered the richest of the former Soviet republics in oil and gas resources, with more than 60bn barrels, according to industry analysts. Azerbaijan, a significant source of oil for more than a century, has the geophysical potential for much greater production than Soviet geologists thought. Turkmenistan, whose major resource wealth is natural gas, ranks third among the regional states in estimated oil reserves, although a recent domestic report, based on up-to-date geological studies, claims that the country's reserves may be as high as 46bn barrels.[14] Uzbekistan, with its 230 known oil- and gas fields, and Georgia also possess energy reserves, though not on the scale of their neighbours.[15] Armenia, Kyrgyzstan and Tajikistan are thought to have minimal deposits.

It is therefore ironic that, newly independent, and possessing higher-than-expected oil reserves, all the relevant oil-producing countries of the region, except Uzbekistan, have experienced declining production levels against a background of poorly performing economies. This can be attributed to the difficulty in short-term capitalisation on discovered reserves. Further obstructions of regional oil-based economic growth are 'the fragmentation of the All-Union industry, the subsequent breakdown of business links between technologically-interdependent national enterprises, inadequate equipment and material supplies and lack of capital investments ... and huge payment arrears built up by insolvent downstream enterprises'.[16] While the Caucasian and Central Asian states are strong from the point of view of oil potential, they are afflicted by infrastructural weakness leading to slow economic growth, and a degree of international vulnerability.

12

II. OIL EXPLOITATION AND EXPORT

Conflicting interests of the Caucasian and Central Asian states, the dispute over demarcation of the Caspian Sea, domestic political instabilities, and technical factors play important roles in Caspian oil development and export. In addition, external interests, based on commercial, domestic and international policies, create a range of pressures on the oil development and strategic formation of the region.

External Political Players
As one of the most turbulent areas of the world, the history of Caucasian and Central Asian states has been one of regional influence, political manoeuvring, shifting alliances, commercial competition and outright war. The region has, in part as a result, been vulnerable to the interventions of a variety of interested parties outside the region. Russia, Turkey, the UK and Iran have, at different times, controlled substantial areas, seeking either to exploit the region's natural endowments, or to use it as a launching point from which to conquer India and the rest of South Asia. On the few occasions in the past two centuries when the Caspian Basin countries have tried to declare and maintain some kind of independence, they were met by failure and eventual submission to one or another of the imperial powers.

With the dissolution of the Soviet Union, these countries are again trying to establish independence. Kazakhstan, Turkmenistan and Azerbaijan, in particular, see the key to that independence in their abundant energy reserves. One of the main problems, however, is that even if these landlocked countries were able to overcome the enormous problems of getting the oil out of the ground, they would not be able to export it without the cooperation of their neighbours. This leads, at the regional and international levels, to a complex series of manoeuvres and kaleidoscopic alliances and counter-alliances, designed to gain access to, and influence over, some of the most valuable resources in the world.

Russia
In the course of the Great Game, Russia gained control of both the Caucasus and Central Asia. This control was maintained and reinforced during the Soviet period, and the ties Moscow retains with the region's states, as a result, remain strong. The break-up of the Union, however, left Russia with a set of new and theoretically self-

determining states in its backyard, forming a cordon between it, China, Turkey and the broader Islamic world. This required the evolution of a new strategic regional policy. That wider policy has been, and remains, inextricably bound to Russia's position on oil. However, while Moscow continues to articulate strategic interests in the area, these interests do not seem to cohere in an organised or disciplined 'grand plan'.

President Boris Yeltsin and other Russian officials have used a number of formulations to articulate their country's special 'responsibility' for the former Soviet Caucasian and Central Asian region. The impact of a more assertive Russian foreign policy has been felt in the past two years in Georgia, Nagorno-Karabakh, Azerbaijan and Kazakhstan. Shortly after independence, Georgia attempted to take a more autonomous path by seeking closer relations with countries such as Turkey and the US, and by bringing in international bodies – the United Nations and the Conference for Security and Cooperation in Europe (CSCE, now the OSCE) – as a neutral broker for its internal conflicts. However, internal political instability continues to allow Georgia's neighbours and former Muscovite masters to exert significant influence over it, rather than allowing it to develop a sovereign stance.[1] Georgian President Eduard Shevardnadze seems now to have realised that some form of accommodation with Russia is necessary.[2]

In Nagorno-Karabakh – an ethnic Armenian enclave in the middle of Azerbaijan – the context of the Armenian–Azerbaijani conflict that began in 1988 is highly suggestive of Russia's emergent policy towards the Caucasus and Central Asia, designed to maximise Moscow's role in ending the war. In early attempts at mediation by the OSCE, Russia refused to concede the Organisation a dominant role in the peace process. Only in December 1994, at the Budapest Summit, did Russia agree that the OSCE should take the lead in seeking a peaceful settlement in Nagorno-Karabakh. Even then, Moscow tried to persuade the OSCE and/or the UN to sanction and finance Russian peacekeeping troops to monitor the resolution of the crisis. (Azerbaijan is one of the few former Soviet republics without Russian troops on its soil and does not wish to see those troops return.) Analysts have noted that the war, which has rendered some non-Russian pipeline routes unusable, gives Russia an advantage in securing its aim for Caspian oil to be exported only through Russian domestic pipelines.[3]

Some regional specialists and Caucasian officials believe that Russia was involved in the downfall of former Azerbaijani President

14

Abulfaz Elchibey after he adopted a strong pro-Turkish and anti-Russian stance that Moscow saw as a direct threat to its influence in the region.[4] In June 1993, just as Elchibey was preparing to sign an agreement with international oil companies, and showed signs of willingness to move towards a resolution on Nagorno-Karabakh with the CSCE, he was ousted in a coup that brought Gaidar Aliyev to power.[5] When Aliyev also began to assert a more independent policy, although not as blatantly anti-Russian as Elchibey's, Azerbaijanis organised another coup attempt, thought to be connected with Moscow.[6]

In Kazakhstan too, Russia has brought economic and political pressure to bear on Nursultan Nazarbayev's government. It has on occasion halted payment for coal mined in Kazakhstan's Karaganda region, obstructing capital flow to the new state, and has reduced Kazakhstan's fuel supplies.[7] In stopping coal payments, Moscow may have been trying to exacerbate already existing sensitivities between Russian coal miners and the Kazakhstani government.[8] In another case, a US oil executive in Almaty complained that 'Russia is holding Kazakhstan hostage' and asserted that 'Moscow's pressure was prolonging delays in a production contract he had been trying to close for several months.'[9] 'By blocking and delaying new projects, the Russians have managed to win entry into practically every major [energy] deal – with little or no cash investment on their own part.'[10] Like other former Soviet republics, Kazakhstan has been forced to reach accommodation with Russia and, in 1995, signed a series of agreements enhancing Russia's influence, including one that merges some functions of their military institutions.

A measure of success in maintaining post-independence influence in the Caucasus and Central Asia notwithstanding, Russia's current policy towards oil in the region is characterised by two basically contradictory schools of thought. The first has been espoused by present Foreign Minister Yevgeny Primakov and other officials, who interpret Russian policy within a traditional balance-of-power framework. According to scholar Robert Barylski, on 21 July 1994 former Foreign Minister Andrei Kozyrev and Primakov (then KGB Director) convinced Yeltsin to sign a secret directive 'On Protecting the Interests of the Russian Federation in the Caspian Sea', which argues clearly that Russia should maintain its 'sphere of influence' in the region.[11]

This group views oil as a central instrument in maintaining that influence. In terms of international competition for the region's oil, the Primakov school sees development and export of oil in zero-sum terms, rather than as a cooperative effort from which everyone can benefit. The Russian press and those Russian officials subscribing to this view warn against new Western joint ventures in the non-Russian parts of the former Soviet Union. They see the substantial involvement of Turkey, the US, the UK and other Western countries in the Caucasus and Central Asia as a potential erosion of Russia's influence.

The second school, almost diametrically at odds with the first, has been supported by Prime Minister Viktor Chernomyrdin and other oil-industry officials. They welcome Western participation in the development of Caspian oil, as a means of ensuring access to capital and advanced technology. They have worked for Russian inclusion in Western consortia in order to improve their own technology, to establish a foothold in world oil markets, and to share in the profits available on those markets.

These two competing Russian schools are reflected in the often contradictory or fragmented responses of the Russian government to Caucasian and Central Asian issues. For example, when a multi-billion dollar Azerbaijani oil deal was signed in September 1994, with the Russian oil company Lukoil as a signatory, the Russian Foreign Ministry held an official news conference to condemn the deal as illegitimate, while a representative of the Russian Energy Ministry participated in the signing ceremony. That said, the general trend of Russian policy in the area has been to strengthen and consolidate influence.

High-level Russian officials have stated publicly that since Russia developed the region's existing infrastructure under the Soviet Union, it owes Russia a debt for this service.[12] Moscow has bid aggressively for a stake in lucrative oil deals, and was successful in securing a 10% share of an Azerbaijani consortium for its domestic oil company, Lukoil, and a 15% share in the Kazakhstani Karachaganak deal for Gazprom. Industry sources also believe that Russia is pushing hard for a share of Tengizchevroil, one of the former Soviet Union's largest joint ventures with a potential investment of $20bn. In addition, Russia has exercised leverage over companies in the Caspian region by putting pressure on their holdings in Russia.[13]

With control over the only major export pipeline for this oil, Russia has been able to put pressure on Kazakhstan, for example, to cooperate on the proposed construction of a second pipeline, also planned to run through Russia. In June 1994, Kazakhstan openly accused Russia of decreasing the volume of Kazakh oil through the existing line to force Kazakhstan to accede to Russian political and economic demands.[14] The present pipeline does not have the capacity to handle the additional oil that will come from Kazakhstan and Azerbaijan, and if another has to be built, Russia would like it to pass through Russian territory.

However, the second pipeline proposal is complicated by a number of factors. Foremost among these is resistance by most of the newly independent states to sustained Russian control of export routes. Furthermore, with the continuing Chechen conflict, Russia has to concede that the region – and any pipeline crossing it – is vulnerable to the threat of further unrest and terrorism. Finally, Turkey's review of acceptable levels of tanker traffic in the Turkish Straits complicates the proposal to build a main export pipeline to Russia's Black Sea ports, since the Straits are the main existing route out of the region to the Mediterranean. Russia is considering various alternatives to the Straits, including an option to send oil through Bulgaria to the Mediterranean.

Overall, it appears that despite profound disagreements in Moscow about the appropriate view to take of Russia's strategic and commercial interests in the Caucasus and Central Asia, and correspondingly fragmented policy implementations, Russia continues to exercise significant influence over the region. With the increase in oil production from the new oil-rich states, the degree of Russian influence will depend critically on Moscow's developing stance with respect to joint ventures in the area, and the direction in which the export infrastructure develops.

The United States

The US has three main policy goals in the region. The first is support for the sovereignty and independence of the countries of the region. The US takes the view that oil is the key to the economic viability of several of these countries, particularly Azerbaijan and Kazakhstan, and that oil development in those two could also bring benefits to others, such as Georgia and Armenia, depending on export routes.

Second, the US supports its own commercial involvement in the region's oil production and export, on the basis that its domestic companies' involvement can help to further economic reform and facilitate the region's entry into the world economic market. Such commercial involvement could also enhance the US presence in the Caucasus and Central Asia, and in developing a highly valuable resource to which private companies bring necessary capital, management and technology. Finally, it is hoped that the involvement of

US companies in successful and lucrative oil deals will bring economic benefits to the US.

Third, US policy supports the diversification of world oil supplies to reduce future dependence on Persian Gulf oil. This is considered particularly important in the run-up to and after the year 2000 during which time, according to some projections, world oil capacity will not keep pace with the demand created by economic growth. This is not to say that the world will experience the same oil shocks that occurred in the 1970s, but the margin between production and demand may be narrower than it is now, as some present resources dry up. Caspian oil will not begin to make a significant difference until after 2005, but the US endorses as early a start as possible to planning and development.

One challenge confronting Washington is balancing commercial interests in the region with other interests and foreign policy goals. These include a desire to contain Iran because of its use of terrorism to derail the Middle East peace process, and its quest for weapons of mass destruction; an interest in encouraging Russian political and economic reform and fair commercial practices in the region; support for an end to regional conflicts, including in Nagorno-Karabakh, Chechnya and Georgia; and the desire to maintain a good relationship with Turkey, a critical ally in an area that is of top national-security interest.

The rapid development of oil projects is crucial for the Caucasus and Central Asia, whose countries tend to be poor, and economically dependent on Russia. US oil policy in the Caspian region – designed to support domestic company involvement in the region's projects; to bolster the states' independence and sovereignty; and to promote the development of internationally accepted business practices in the former Soviet Union – relies on four main instruments:

- *Active diplomatic support* at all levels, from embassy officers to the President. The President, Vice-President and several cabinet members have worked actively to pursue US goals in a number of high-level meetings with all countries involved – particularly the $20bn Tengizchevroil project in Kazakhstan, the Azerbaijan international oil consortium, and in the Caspian demarcation issue. US officials have maintained extensive contacts with domestic company representatives in order to coordinate strategies for the promotion of national business interests.[15] President Bill Clinton's October 1995 discussion with Azerbaijani President Aliyev is thought to have been a decisive factor in the latter's agreement to both northern and western routes for short-term export. Vice-President Al Gore has also been active in promoting US oil policy in the region through contact with regional leaders and through the special framework for cooperation with Russia – the Gore–Chernomyrdin meetings. He has also been pushing an environmental protection agenda, encouraging oil projects that are safer for the environment, and securing the cooperation of US companies. For example, the AIOC's first production-sharing agreement includes much higher standards for environmental protection than any used before in the Caspian.[16]

- *Government trade and commercial bodies*, including the Overseas Private Investment Corporation, the US Department of Commerce, the Export–Import (Ex–Im) Bank and the Trade and Development Agency. These are either already involved in projects, or examining ways to assist Caspian states in getting their projects started more quickly and efficiently. For example, Ex–Im Bank signed a Project Incentive Agreement with Kazakhstan on 27 April 1994, allowing the Bank to engage in limited recourse financing in all sectors including oil.

- *Substantial technical assistance* to help these countries develop their legal and commercial infrastructures to meet modern needs and facilitate oil development and export projects.

- *Support for International Financial Institutions'* (IFI) efforts at institution-building and infrastructure policies in these countries.

Given these policy goals and instruments, the US has established parameters to underpin its policy. Future national political and commercial decisions will be made within these parameters as the situation develops. They include:

- *Multiple short- and long-term routes.* The US has promoted this policy since 1994 because it encourages commercial competition, keeping tariff rates lower, safeguarding exports against interruption by avoiding dependence on a single route, and endorsing fairer commercial practices.

- *A route through Turkey (as one of several routes).* This will: augment the total amount of pipeline capacity to export oil from the Caspian region, relieving current pressures on the Russian pipeline system; decrease Caspian countries' dependence on a single route through Russia; allow exporters to avoid the weather and capacity problems at the Russian port of Novorossiysk; reduce the potential for oil spills and tanker accidents in the Black Sea and the Turkish Straits; and reduce the pressure for a route through Iran to the Persian Gulf.

- *Opposition to projects that give Iran significant political, material and economic benefits.* The US has encouraged Caspian countries to minimise Iranian involvement in oil projects as part of an overall effort to contain Iran. This is a result of Iran's attempts to jeopardise the Middle East peace process and its use of terrorism as a foreign-policy tool.
- *Restructuring the Caspian Pipeline Consortium project* to allow it to obtain support of the international financial community.

Oil companies and the Caucasian and Central Asian countries are in the process of sorting through proposals for medium- and long-term exports, including feasibility studies, cost estimates and risk assessments. The US has been careful not to push ahead of the commercial game, particularly since it is not providing financial backing. The parameters described above define the policy as far as possible, in the context of developments to date. As the situation evolves, US policy-makers will further refine those parameters in a way that best suits their national interests.

For example, companies and countries involved in the Azerbaijani and Kazakhstani oil projects are presently examining possible pipeline routes through either Georgia or Armenia. The US would consider supporting the future development of these routes depending on whether the projects can attract suitable financing and are supported by commercial interests. A route through Armenia, under the right conditions, could enhance the Nagorno-Karabakh peace process, if the parties could agree on a partnership arrangement, and if the conflict settlement process had progressed sufficiently to give investors confidence in the project's viability. The US has been an active participant in seeking a resolution to the seven-year-old war, and if the peace process advances far enough, the US can find ways to encourage financing.

A factor that negatively affects US efforts to mediate in the Nagorno-Karabakh conflict, and Washington's ability to become involved in the development of Azerbaijan's strategic oil sector, is the Freedom Support Act (Section 907), which restricts all aid, including humanitarian, to Azerbaijan. Both the Bush and Clinton administrations have objected to Section 907. Fundamentally biased by the assumption that only one side of the Nagorno-Karabakh conflict is to blame, Section 907 prevents the US government from, among other things, helping Azerbaijan improve and shape its legal and commercial infrastructure – a move that would facilitate US oil company efforts to develop this strategic resource.[17] Section 907 prevents the US from offering advice on democratisation and economic reform, effectively bans humanitarian assistance for civilian victims of the war, and constrains the flexibility of US negotiators in crafting solutions with incentives for both sides.[18]

In late 1995, Texas Congressman Charlie Wilson added language to the House Appropriations Bill that would allow the US to provide humanitarian assistance to Azerbaijan if the President determined that non-governmental organisations (NGOs) were not providing adequate support for refugees and displaced persons. Although this will improve the situation and allow some forward movement on the issue, US policy will still be biased and the remaining restrictions will inhibit its effectiveness in the region.

Turkey
Turkey has periodically had strong ties to the Caspian region, particularly to Turkic ethnic groups there. These ties, however, were considerably weakened during the Soviet period. The demise of the Soviet Union has given Turkey an opportunity to renew its historical association with its ethnic groups in the Caucasus and Central Asia, and to increase once more its influence in the region. This has been established as a priority for Turkish foreign policy. Ankara's traditional relations with the Caspian region have often been nurtured at the expense of Turkey's historical rivals, Russia and Iran. In an attempt to profit economically from new investment opportunities, Turkey has presented itself as a development model for the newly independent states, and has generally been seen as a more attractive example to emulate than Iran which, equally keen to play a leading role in the area, has also offered the new states advice. Turkey has concluded a number of political, military and economic agreements with the Caucasian and Central Asian states, and commercial ties have deepened, particularly since 1991.

Turkey's initially enthusiastic programme to further relations with the region has been tempered by the realisation that change will come more slowly than had been hoped, and that some Soviet habits die hard – particularly in the economic realm. Turkish policy-makers have consequently adopted a more long-term view, based on modified expectations about what can be accomplished in the short term. Having said that, Turkey remains committed to a close relationship with the countries of the Caspian Basin for both political and economic reasons.

Although it is understandable that Turkey has concentrated on its relations with fellow Turkic peoples in the region, Ankara has also been working quietly to forge better relations with its historical enemy, Armenia. The latter's President Levon Ter-Petrosyan recognised that an important key to the survival of Armenian independence was good relations with all its neighbours, even Turkey. For its part, Turkey clearly appreciates the strategic importance of Armenia, and recognises that there are benefits to a stable Turkish–Armenian relationship. Although both governments have come under fire from their extreme nationalists, both have continued making progress by gradually increasing official contacts and loosening restrictions on cross-border trade – although progress has been unpredictable over the past few years because of the Nagorno-Karabakh conflict.

Turkey has made significant efforts in forming sound relations with its closest neighbour, Azerbaijan, and with the most populous of the Central Asian states, Uzbekistan. The presidency of Azerbaijan's Abulfaz Elchibey, from June 1992 to June 1993, marked a high point of Turkish influence in the region. Although it is arguable that Elchibey was overthrown for internal and external reasons – and maybe even with support from Moscow – Azerbaijan's overt move towards Turkey clearly worried many in Moscow. Turkey has expressed concerns about Russia's behaviour in the Caucasus and Central Asia, and has emphasised the need for these countries to decrease their dependence on Moscow, particularly for an oil export route. The Turks point to Russia's policy on the Nagorno-Karabakh issue, and the Abkhaz separatist rebellion in Georgia as prime examples of what they see as Russia's aggressive regional posture.[19]

Underpinning its mediation between Russia and the Caucasus and Central Asia is Turkey's principal interest in increasing its own influence there. Ankara is particularly keen to build a pipeline to carry Caspian oil out through Turkey, anticipating the substantial benefits in terms of income and jobs such a pipeline would bring. However, the outcome of Turkey's review of limits on tanker traffic through the Bosphorus Straits – for reasons including safety and the environment – could significantly affect wider deliberations on transport routes for this oil.[20] Plans for a pipeline through Turkey may also be complicated by Ankara's struggle with Kurdish separatists in the region through which the pipeline would pass. That said, a decline in Kurdish terrorism during the past two years, and the

Turkish government's pledge to provide protection for the pipeline, may help ease concerns on this point.

Another, perhaps less tractable, problem for Turkey's pipeline aspirations is that oil from Azerbaijan and/or Kazakhstan would have to go through Iran and/or war-torn Armenia or a politically precarious Georgia before reaching Turkey. None of these routes is particularly secure; all pass through politically unstable regions. Further, financing the infrastructural improvements required for an Iranian route would present serious policy concerns for the US. These routes will be examined in detail later in this paper.

Iran

Like other traditional regional powers, Iran has complex historical ties with the Caucasus and Central Asia. For example, an estimated 16 million ethnic Azeris live in northern Iran – nearly twice as many as in Azerbaijan. Indeed, during former Azerbaijani leader Elchibey's presidency, these cross-border ethnic links were a source of tension in Azerbaijani–Iranian relations, since Elchibey openly advocated the unification of Iran's Azeris with Azerbaijan.

Tehran has not been slow to take advantage of opportunities in the region offered by the disintegration of the Soviet Union. From 1991, the Iranian government recognised the independence of all the new states, with the exception of Moldova. Specific Iranian goals in the region include political influence, profitable economic and commercial relations, the spread of religious ideology, procurement of former Soviet weaponry, and the acquisition of nuclear expertise and materials. At a more profound level, Iran is interested in showing that it can take on the mantle of regional leader. Iran's President Hashemi Rafsanjani, referring to his country's role in the Caucasus and Central Asia, frequently talks of its emergence as an 'economic trade center'.[21] Critically, as Iran's own oil resources become depleted, its geographic position offers opportunities for influence over the region's new reserves.

Iranian officials, religious leaders and businesspeople have visited all the countries of the region, increasing contacts with the new governments and their people. Although Iran does not have the economic capacity to offer significant aid, it has pursued with great zeal wider economic contacts and joint ventures, particularly in the area of oil and gas. While establishing relations with all the region's states, including Christian Armenia, Iran retains its closest ties with Turkmenistan – alarming Turkmenistan's Central Asian neighbours, particularly Uzbekistan and Kazakhstan, both wary of Iran's role in supporting the armed opposition in Tajikistan.[22] This wariness notwithstanding, other Caspian Basin countries maintain ties with Iran, in part to balance their relationships with Russia and Turkey.

Iranian Foreign Minister Ali Akbar Velayati has frequently talked of the 'revival of the silk route' during his visits to the region, and has pushed for the rapid construction of rail and road links between Iran and the Caucasus and Central Asia.[23] Such links would include a rail line connecting Iran with Turkmenistan, and the improvement of transportation links with Azerbaijan. If completed, they would indeed form the basis of a silk route revival, because most of the countries of the Caucasus and Central Asia would have transportation links through Iran to the Persian Gulf. In the longer term, Iran has designs on a link through Turkmenistan and Kazakhstan to China.[24]

As part of this silk route revival, Iran has sought shares in a number of Caspian oil and gas development and export ventures, and is aggressively pushing these states to explore both short- and long-term export arrangements as an alternative both to the Russian pipelines, and to the unstable areas of Nagorno-Karabakh and Georgia that any Turkish route would involve. One possible arrangement would be a system of swaps, in which oil is sent to northern Iran, and Iran sends out to the Persian Gulf oil of equivalent value. Such an arrangement, on a limited basis, appeals to some of the region's countries, because it offers an alternative to a complete Russian monopoly over exports, allowing slightly more room for manoeuvre in oil trade negotiations. In the long run, however, the value of Caspian oil depends somewhat on its capacity for delivery to Western markets without dependence on the crowded Persian Gulf and Suez Canal routes.

Like others, Iran has used a combination of incentives and threats, particularly with Azerbaijan, to try to achieve agreement on a pipeline route and on other areas of oil trade. In the context of limited success here, Iran failed, however, to secure a share in the lucrative $7–8bn Azerbaijani international oil consortium deal, primarily due to US opposition to Iranian participation. Industry rumour – although no substantiating agreement has been signed – suggests that Azerbaijan intends to give Iran a share in this or another oil-field to smooth the latter's ruffled feathers, and to enlist Iranian support for Azerbaijan's view of the Caspian boundary dispute.[25] Elsewhere, Iran has encouraged Kazakhstan to send more of its oil through to Iranian ports in swap deals.

Given its wider political isolation, Tehran has become increasingly active in trying to amplify economic and political contact with the Caucasian and Central Asian states. In February 1992, Iran hosted what was considered a successful meeting of the Economic Cooperation Organisation (ECO) – Iran, Turkey, Pakistan, Azerbaijan, Turkmenistan, Uzbekistan, Tajikistan and Kyrgyzstan with Kazakhstan as an observer – to discuss Caspian issues, and to try to take a regional leadership role in economic, particularly oil and gas, affairs. ECO agreed to set up a Caspian Cooperation Organisation (CCO) with its headquarters in Tehran.[26] At an international conference on Caspian cooperation in Tehran in April 1992, the participants, including the Caspian littoral states, signed a protocol 'on the protection of the Caspian environment, navigation, passenger traffic between their respective ports, and related issues'.[27]

Iran and Turkmenistan have signed a number of cooperation agreements during the past two years in oil- and gas-related ventures, including a large-scale gas pipeline proposal that would take natural gas from Turkmenistan through Iran to Turkey and then on to Europe. Former US Secretary of State Alexander Haig is involved in coordinating this pipeline deal. According to industry sources, however, the deal is in considerable trouble, most seriously because it is unable to attract significant financing.[28] In April 1994, Turkmenistan also concluded an agreement with Iran on the construction of a pipeline that would supply 6m tons of Turkmen oil to northern Iran.[29]

Iran has sought economic and commercial cooperation at various times with Turkey, Russia and some of the Western European countries involved in the Caspian region, but generally these are alliances of convenience and tend not to be long-lasting. Iran has numerous ties with Russia, including significant economic and trade relations on which Iran depends. At the same time, Iran is in competition with Russia throughout the region, particularly on short- and long-term export issues. For example, Iran has offered to cooperate with Russia on the Caspian boundary dispute, in part as an accretion of leverage to pressurise the Caspian littoral states to work with Iran on other issues. On the other hand, Iran itself has been subject to Russian leverage over the trade Tehran desires, particularly in the areas of arms and civilian nuclear power-plant material. As a result, Iran has shifted its position on the Caspian boundary issue several times, depending on the particular constellation of influencing factors and Iranian needs at the time.

Iran's efforts to sabotage the Middle East peace process, its use of terrorism as a foreign-policy tool, and its pursuit of weapons of mass destruction have led the United States to take steps to put economic pressure on it. As long as Iran pursues its destabilising activities, obtaining finance for any venture will be difficult. Although other Western countries have not adopted policies as tough as those of the United States, the US has exerted strong pressure on other countries to punish Iran for its behaviour, successfully arguing for a block on sensitive technology transfers, and denying Iran access to international finance. Since President Clinton announced a US trade embargo in mid-1995, no government or international financial institution has extended new official credits to Iran. The US has also been effective in its efforts to limit foreign investment in the development of Iran's petroleum resources.

China

Like Turkey and Iran, China is also a possible model for the countries of the Caucasus and Central Asia. The Chinese model holds some appeal for the region's more conservative and authoritarian leaders, who wish to reform economically without doing so politically. Uzbekistan's President Islam Karimov and his government often cite the virtues of China's reform effort, claiming that this might serve as a better model than that of the West. Relations between China and the Caucasian and Central Asian countries have been basically positive, including official visits, protocols and agreements. Some tension with Kazakhstan exists over Chinese nuclear testing at Lop Nor.[30] But according to scholar Ross Munro, trade between China's north-western Xinjiang Autonomous Region and Kazakhstan is responsible for much, if not most, of Kazakhstan's economic growth in 1992.[31] Uzbekistan and Kyrgyzstan have sought closer relations with Beijing, and China has become one of their largest trading partners outside the former Soviet Union. Munro asserts that it may be 'conceivable that Kyrgyzstan's economy will be dominated by China before the end of the decade'.[32]

China and Central Asia have several ethnic groups in common – Kazakhs, Uighurs, Tatars, Tajiks and others. Deep antagonism between the Han Chinese and these groups characterise the history China's relations with its neighbours to the north and north-west. In recent years, Beijing has allowed greater contact between Central Asians and their ethnic counterparts on the Chinese side of the border, despite these ethnic tensions. The Chinese and the Kazakhs quietly agreed in the early 1990s to allow thousands of ethnic Kazakhs to emigrate from Xinjiang to Kazakhstan.[33] Some see this as part of a Chinese policy to change the ethnic balance of Xinjiang in favour of the Han, making it easier for Beijing to retain control there.[34]

China became a net oil importer for the first time in its history in 1993. According to economist Mamdouh Salameh:

In the face of declining oil reserves, flagging output and rising domestic consumption, China must make significant new oil discoveries if it is to maintain the momentum of its economic growth and avoid becoming heavily dependent on oil imports in the late 1990s and beyond.[35]

Some Western geologists believe that China's Tarim Basin, near the border with the former Soviet Union, holds oil-fields that may be even larger than those in Kazakhstan. In the absence of detailed studies, however, such claims are tentative at best. Even if the Tarim Basin does have significant oil reserves, developers would face harsh geophysical conditions in shifting sands that reach up to 70 storeys high while having to drill some of the deepest wells in the world.[36]

China has recently invited foreign companies to help explore and develop the region,[37] reserving, however, the most valuable and promising fields for its own State firms, even though it lacks the capital and technology to work them.[38] If hopes for the Tarim Basin bear fruit, Beijing may become more interested in participating in one or more of the many pipeline proposals under discussion. And if Tarim resources are sufficiently large to develop but not sizeable enough to warrant a long pipeline, China could send the oil by rail to eastern Kazakh refineries.

In the long run, if Kazakhstan oil projects continue to develop, and the political and economic conditions are favourable, China may be interested in linking up with a possible West–East pipeline taking Kazakh and Chinese oil to the Far East, and involving the participation of Korea and Japan.[39] A proposal like this, however, would not be realistic for at least another decade, and it is very difficult to predict the political circumstances that would affect the realisation of such a plan. Any oil or gas pipeline linking China with Central Asia and the Caucasus would be very long and costly, traversing vast stretches of difficult terrain. Although this would not necessarily preclude constructing this line, such factors would certainly be a major consideration.

Having said that, Turkmen President Saparmyrat Niyazov began discussions in 1992 with Beijing and the Japanese company Mitsubishi for a gas pipeline to China.[40] Also, during a meeting on 12 September 1995, Kazakhstani President Nazarbayev and Chinese President Jiang Zemin agreed to study possible oil pipelines connecting western Kazakhstan with the eastern coast of China.[41] In

1993–94, several US businesses also expressed interest in the project, but as of early 1996, the proposal has still not progressed much beyond the feasibility study stage.

Pakistan
Although Pakistani leaders talk as nostalgically as the Iranians about the recreation of a silk route, they confront problems of chronic unrest in Afghanistan which lies between Pakistan and the Caucasus and Central Asia. Financing a venture in silk-route-style transport links would be difficult because of this instability. In addition, the Pakistanis must deal with problems in their own country and the problems of management at their port in Karachi. On the other hand, Pakistani and Afghan problems, although complex, are not necessarily any less tractable than those on some of the routes that would run through the Caucasus. On 21 October 1995, the US Unocal Corporation and the Delta Oil Company of Saudi Arabia signed a protocol of intent for a proposed oil pipeline to extend from Turkmenistan through Afghanistan to Pakistan. In signing the protocol, participants admitted that such a project might be difficult to finance right now, but they expressed confidence that it might be possible in the future. Pakistani routes, however, have not attracted the same level of commercial interest as the other routes.

Other Western and Developed Countries and Institutions
Other Western and developed countries – the UK, France, Japan, and Italy – share policy goals with the US, including encouraging stable, independent, secular, democratic and market-oriented countries in the Caspian region, and pursuing profitable commercial deals that would benefit their own domestic firms. They have consequently established political, aid and trade programmes with the region's countries similar to those of the US. As well as this political and aid cooperation, however, there is also intense commercial competition for shares in deals such as the Azerbaijani consortium, which give rise to differences among the Western countries.

In spring 1995, both France's President and Prime Minister contacted President Aliyev in an ultimately unsuccessful attempt to increase Elf-Aquitaine's share in the Azerbaijani consortium over those of the US companies Exxon and Mobil and Italy's Agip. And differences between the US and others in the West arise over Iran. Unlike some Western European countries, the US advocates a policy of economic pressure against Iran to change its behaviour. Again, unlike the US, Western countries have not been explicit in their public policy pronouncements on Caspian oil issues, particularly on the pipeline proposals.

Like Western countries, international organisations such as the World Bank and the European Bank for Reconstruction and Development (EBRD) may play a greater role in future in the region's oil and gas ventures. The World Bank has already initiated projects to help several Caspian countries develop their energy infrastructures and legal frameworks. In early 1995, it granted Azerbaijan and Kazakhstan technical assistance loans of several million dollars to develop their management of oil resources.

Other organisations such as the European Union (EU) may also increase their participation in the future. The EU already provides a large amount of humanitarian and technical assistance to the Caucasus and Central Asia, but thus far has not provided assistance in the oil and gas sector. It is, however, examining proposals for a Greek, Russian and Bulgarian bypass pipeline to take oil from the Bulgarian Black Sea port of Burgas to the Greek Mediterranean port of Alexandroúpolis. If the project goes ahead, it will increase the amount of oil that can be shipped through the Black Sea, affecting overall export considerations from the Caucasus and Central Asia.

The Caspian Legal Regime Dispute
The dispute over the demarcation of the Caspian Sea is a good illustration of the interplay among the countries in the region. Before the break-up of the Soviet Union, the legal status of the Caspian Sea was established under Soviet–Iranian treaties signed in 1921 and 1940. In 1992 and 1993, after the break-up, the Caspian littoral states met in Astrakhan and Tehran to discuss Caspian demarcation and other Caspian-related issues such as regional development, the environment, and fishing rights. On several occasions, all the states except Russia agreed to a plan, formulated by Kazakhstan and based on the old Soviet republic boundaries, to divide the Caspian into individual sectors bounded by equidistant lines from the shores of the bordering states. This was the *de facto* arrangement the new states had been using, and one which is consistent with the way other similar bodies of water are legally divided. Under such an arrangement few, if any, substantial resources would be in Russia's sector and certain groups in Russia, including the Foreign Ministry, announced their objection to any division of the Caspian Sea along state boundary lines. In November 1993, however, Russian Fuel and Energy Minister Yuri Shafranik signed an agreement with Azerbaijan that recognised an Azerbaijani sector in the Caspian Sea.

Other Russian officials began to take a harder line on the Caspian, particularly in 1994 when Russia adopted a more aggressive stance in several areas of the former Soviet Union, including the Caucasus, Central Asia and Ukraine. In spring 1994, the Russian Foreign Ministry and elements in Yeltsin's staff called for the classification of the Caspian as a condominium with no sectoral divisions. This would mean that each littoral state could claim only resources lying within 10 nautical miles of its shores, the rest being shared equally among all littoral countries.[42] Some legal experts believe that Russia would not be able to prove in an international court of law a condominium case for the Caspian rather than a sectoral division as is common for such bodies of water. They admit, however, that while a negative outcome for Russia is likely, it would not necessarily be the result of adjudication.[43]

Russian politicians and officials have issued public warnings that opening up the Caspian Sea to international oil and gas developments over which Russia had no control would erode Moscow's security and its political and economic influence, damaging Russian interests in the region. As more and more lucrative oil and gas deals with Western companies approached closure in the Caspian area, Russian Deputy-Premier Aleksander Shokin and Foreign Minister Andrei Kozyrev took a stronger stance, in particular by invoking the unresolved Caspian demarcation issue to oppose the Azerbaijani international consortium. In October 1994, Russia circulated a paper at the UN warning that Moscow 'reserved the right to take appropriate measures' against Caspian states that unilaterally begin exploring the Caspian seabed.[44]

Moscow officials have also expressed concern about protecting the sea's environment – important given that the Caspian is the source of almost 90% of the world's caviar. These concerns, however, have a hollow ring. First, because the Azeri project's Production-Sharing Agreement included more environmental safeguards than any project ever implemented in the Caspian; and second because long-term Soviet and Russian disregard for the environment elsewhere has led to dire consequences in places such as Komi.[45]

For their part, the ex-Soviet Caspian littoral states would like to resolve the legal regime soon, although they are not prepared to relinquish what they consider to be their rights. Many are concerned that Russia's stance will scare away potential foreign investors needed to fund large oil and gas projects. These projects are crucial for the survival of local economies. Azerbaijan has taken the strongest stand among the littoral states against Russia's efforts to establish a condominium division of the Caspian, mainly because this would affect its immediate interests.

In October 1994, Turkmen President Niyazov publicly supported Azerbaijan's right to develop its Caspian sector.[46] Turkmenistan does not intend to develop its own Caspian resources in the near future, so the Turkmen position in this respect may vary over time. Kazakhstan, on the other hand, has been trying to find an accommodation with Moscow on the Caspian issue so that it can move forward with much-needed development projects. At the beginning of 1995, Kazakhstan's President Nazarbayev claimed to have resolved the dispute with Yeltsin, but in 1996 Russia still appears to be pressing its own view.[47]

In November 1994, the littoral states established a Caspian coordinating committee to work on demarcation and other related issues, including navigation and fishing rights. In February 1995, all the states except Azerbaijan agreed to a 20-nautical-mile coastal border for exclusive fishing rights. Azerbaijan did not want to sign the accord because it feared establishing legal groundwork for the condominium arrangement.

In mid-1995, in an attempt to promote a northern pipeline route (as opposed to a western route through Turkey), Moscow softened its more confrontational positions on Caspian demarcation. In late 1995, Russia played down the Caspian issue and Ministry of Foreign Affairs officials made positive noises about resolving the issue through negotiation. On 20 September, Russia did not even attend a demarcation-related meeting of the littoral states in Almaty, although Energy Minister Shafranik went to Almaty a few days later for talks with Nazarbayev about energy issues.

As mentioned above, Iran's past support for Russia has fluctuated according to the calculation of its interests at a given moment. It has been in Iran's economic interest to cooperate with Russia over Caspian demarcation, both because it has an important economic relationship with Moscow and because, like Russia, Iran's potential sector of the sea is not thought to be rich in resources. However, Tehran's interests here as elsewhere are conflicting, and industry analysts point to negotiations between Azerbaijan and Iran to develop some of the former's Caspian oil resources as Baku's way of trying to placate Iran after its exclusion from the Azerbaijani oil consortium and gain its cooperation.[48] Since those negotiations increased in intensity in mid-to-late 1995, Iran has been more or less silent in public on the demarcation issue.

Political Factors Inside The Region

The main political factors inside the Caucasus and Central Asia affecting the development and export of oil are: endemic political corruption; organised crime; political instability within countries such as Azerbaijan and Georgia; a high turnover of Kazakhstani government officials dealing with oil issues; the Nagorno-Karabakh conflict; and the war in Chechnya. The Caucasus (Azerbaijan, Georgia and Armenia), in particular, is very unstable, greatly complicating oil projects in the area. In the Caucasus – including the Russian sector between the Caspian and the Black Sea – at least eight areas are suffering sporadic violence or continuing conflict, and at least four others are potentially unstable. Azerbaijan, for example, has experienced three changes in leadership and numerous attempted coups during the past four years. Oil companies involved in negotiations for the international consortium have dealt with a succession of political players, new governments and new negotiators.

Georgia, which recently signed an agreement for a pipeline route through its territory, has been even more unstable than Azerbaijan, with at least three – temporarily quiescent – disputes among the various ethnic groups in the small country. Georgian President Shevardnadze is only slowly widening his control over the country whose power until recently has rested in the hands of warlords and their private militias. That said, in 1996 Georgia has become more stable – indeed sufficiently so to merit $203m in International Monetary Fund (IMF) loans, and at least $75m in World Bank credits. Shevardnadze has become personally involved in trying to reassure potential pipeline investors, and has offered his own guards to protect a Georgian pipeline. However, the possibility of further Russian pressure, the large number of refugees from Abkhazia, corruption and organised crime combine to make possible financiers tread carefully.

Disorganisation and mismanagement in the governments of the new states is common. For example, early in 1995, one of the managers of a western Kazakhstan government oil project contacted a Kazakhstani/Western oil company joint venture to say that his project would not be meeting its export quota for the month, and offered the extra pipeline space for the Western company's excess oil. When the Kazakhstani Energy Minister heard about the deal, he stopped it and demanded that the government project fulfil its quota rather than sharing it (even though the offer of pipe space was made to a partially Kazakhstan-owned venture). The government project duly filled its quota by pumping oil from the bottom of the storage tanks, which contained a large quantity of salt and water. This contaminated the line to Samara, angering the Russians, who promptly cut Kazakhstan's export quota. A persistently bureaucratic Soviet mentality of quotas at any cost ruined what would have been a cost-effective and logical solution.[49]

The Nagorno-Karabakh war, and the associated political manoeuvrings of all involved, has had a detrimental effect on oil development and export projects. Although a cease-fire has held since May 1994, Armenia and Azerbaijan are still far from a settlement. The tenuous nature of the cease-fire, deeply embedded ethnic hatreds, and the failure of both governments to control renegade extremist groups in the region would probably make companies and banks balk at the prospect of financing a pipeline venture in the region.

Another factor that has adversely affected oil development and export is the war in Chechnya, which has obstructed Russia's efforts to promote a pipeline through that region. The Grozny oil district in Chechnya has for many years been a key oil and gas pipeline junction and a refining and production centre. Some analysts believe that one of Russia's central motivations in pursuing a military option against Chechnya was the former's interest in an oil pipeline to carry Baku oil through Chechnya.[50]

However, unrest in Chechnya and a physical infrastructure devastated by the military conflict, have arguably undermined rather than strengthened Russia's objective of investing in regional construction. Poor Russian judgment and a lack of foresight and communication in the execution of the Chechnya conflict may also have negatively affected original Russian aims.[51] In the end, Russia's actions in Chechnya have almost certainly ensured years of bitter hatred from the Chechens, and the possibility of terrorist acts against Russian interests in the region. Moscow's military action has profoundly affected the development and export of oil in the region – mostly to Russia's detriment.

Turkey is another influential factor in oil exports from the region. As mentioned previously, terrorism on the part of Kurdish separatists could pose a problem for the security of any pipeline through Turkey. Although oil companies have operated in unstable regions with more severe problems, the Kurdish issue is commonly and specifically identified by those companies as a counter-factor in their decision-making. However, over the past two years, Kurdish terrorism has abated and, in addition, the Turkish government has offered to provide security for a pipeline running through its territory.

Technological and Technical Factors

In addition to the variety of external and internal political factors outlined above that condition and, at times, impede the development of Caspian oil, technological and technical issues also play an important and sometimes decisive role in regional projects. The technological category includes: inadequate Soviet-era equipment and technology; aging Soviet infrastructure; Soviet exploitation methods that damaged oil-fields; and varying qualities of oil produced in countries that may use the same pipeline. Technical problems include the lack of commercial, legal and financial expertise that has rendered most of these countries vulnerable to entrepreneurial exploitation. Finally, the absence of a substantial Western-style legal structure to support and build confidence among investors has caused problems in attracting much-needed capital.

As mentioned above, former Soviet technology was incapable of accurately assessing deep oil and gas deposits, and the introduction of new technology, including three-dimensional seismic studies, greatly affected estimates of resources in the region. Existing Soviet-era equipment and infrastructure are insufficient for many of the new projects, making it necessary, therefore, to import modern equipment, including large, heavy pieces and special drilling equipment that must be shipped along Russia's River Volga and into the Caspian Sea. This control over the Volga transport routes gives Russia additional leverage over Caspian development.

Because some of the oil projects are established on Soviet-era infrastructure, or located in areas that were developed during that period, problems have arisen caused by Soviet development techniques that damaged the oil-fields. Further, existing Soviet-era oil and gas pipelines, often characterised from the outset by poor design, shabby construction, inferior materials and shoddy maintenance, are deteriorating with age. The export of both Kazakhstani and Azerbaijani oil is also complicated by differences in the types of oil each country produces. The quality and value of Kazakhstani oil is higher than that of oil from Azerbaijan, and a combination of the two would lower the value of the Kazakhstani oil. While this last problem can be overcome by conventional means, it simply adds to the many difficulties involved in export projects.

From the point of view of technical infrastructure, the Caucasus and Central Asia is compromised by a lack of commercial and financial expertise. This can leave states and their economies vulnerable to unscrupulous international entrepreneurs. Described by one journalist as 'the carpetbaggers', hundreds of businesspeople have flocked to the region.[52] Players include a former US Secretary of State; a former Dutch car dealer who became an exceedingly successful international oil trader; a Lebanese national wanted for questioning by Lebanese justice authorities investigating the collapse of its second-largest bank; and dozens of high-level former officials from countries throughout Europe.[53] As a result, governments of the former communist republics have sometimes lacked the commercial experience to protect their interests against these entrepreneurs. A case in point is the Caspian Pipeline Consortium (CPC) – the result of a 1992 agreement between Russia, Kazakhstan and Oman to build a pipeline to transport oil from the Caspian region to the Black Sea –which gained exclusive rights to build a pipeline for Kazakhstan.[54]

The lack of clearly designed legal structures to protect investments, and the absence of accepted Western commercial practices, have affected investor confidence in the region. Many of the region's governments have moved quickly to try to create necessary legal bases to boost that confidence, but much work remains to be done in this area and progress throughout the region has been very uneven. In Baku, for example, Western specialists believe that at least 80% of Azerbaijan's existing tax legislation needs to be overhauled.[55] Kazakhstan's proposed oil and gas legislation has gone through at least 20 drafts, but was not passed before its parliament was disbanded in 1995.[56] Some improvements in this area have occurred, however, and the governments, often with international aid funds, are trying to move ahead.

One of the most frustrating experiences for Western companies is a lack of understanding in the Caucasian and Central Asian countries of Western contractual practices. After the break-up of the Soviet Union, a number of governments and new companies in the region signed various documents and contracts, without considering them binding. Many examples exist of newly formed governments signing contracts with more than one company for the same project. In the Soviet Union, agreements and documents were signed and government laws approved, but they often had little real meaning. The Soviet Constitution and the Union Treaty are good cases in point. During the dissolution of the USSR, when Moscow tried desperately to hold it together by invoking the Union Treaty, one Soviet Republic leader explained to this author why he had signed agreements on the draft, even though he never intended to adopt the final version. 'In the Soviet Union, we sign things all the time, but it doesn't mean anything'.[57]

Also, the lack of technical, legal and commercial expertise in these countries, particularly in the immediate aftermath of independence, led some states to sign documents legally surrendering important negotiating powers that they may not have given up had they been aware of the consequences at the time. The Kazakhstani government, for example, signed numerous documents with a variety of entrepreneurs who had forged close, personal relationships with Kazakhstani leaders to gain their trust by providing financial assistance at times of critical need.[58]

Faults in the communications, commercial and financial infrastructures have also adversely affected the development of Caspian oil business. Western-style banking is still in its infancy there, although several Western banks are participating in joint ventures in Kazakhstan – in 1994 Chase Manhattan opened its own joint-venture bank with the Kazakhstan government. Also, the Dutch Bank ABN-Amro along with a semi-private Kazakh bank and the World Bank's private sector lending arm, the International Finance Corporation (IFC), began a joint venture in May 1994.[59] Azerbaijan, however, lags far behind in such facilities.

The former Soviet republics also need to develop new communications systems. Until a few years ago, Kazakhstan had only ten telephone lines to the outside world. In just the past four years, communications have improved rapidly with the entry of Western telecommunications companies, but oil industry officials would like to see more improvements to facilitate business in the region.

III. THE MAJOR PROJECTS

External interests, internal instabilities, technical and infrastructural problems have all complicated, slowed and, in some cases, stalled the development of an oil industry in the Caspian Basin. Nonetheless, at least 80 oil- and gas-related joint ventures are operating in the region currently, with over 30 US companies involved in these and other commercial deals. Most large-scale lucrative energy projects in the former Soviet Union are joint ventures, often with state-owned or semi-private oil and gas enterprises. New exploration projects are proliferating, with mixed results. It is difficult, therefore, to predict the shape of the production landscape in ten years' time.

The five largest current development projects are: the Kazakhstan–Chevron joint venture known as Tengizchevroil; the Azerbaijani offshore oil project consortium; the Azerbaijani Karabakh field in the Caspian; Kazakhstan's Karachaganak oil- and gas field; and a set of projects in Kazakhstan's offshore area.

The Tengizchevroil Joint Venture
Tengizchevroil, a joint venture between the US company Chevron and the Kazakhstani Tengiz oil-field, went into operation in April 1993. At $20bn, it is one of the largest single investments by a US firm in the former Soviet Union. Chevron began negotiating the deal in 1990 before the fall of the Soviet Union, and continued with Kazakhstan after its independence. The Tengiz oil-field is the largest oil discovery in the world since the 1970s, with proven high-quality oil reserves of 6–9bn barrels – to be developed, according to the contract, over a 40-year period. Being of a higher quality, Tengiz oil is worth about 20% more than oil exported from Russia to international markets. The Kazakh oil is, however, located very deep, at high temperature and pressure, with a high hydrogen-sulphide content, making it relatively expensive to produce. Although the Soviet Union had begun to develop the field using Western technology in the late 1980s, production had begun to fall just as the Union disintegrated because the existing Soviet technology was not capable of fully exploiting the reserves.

The Tengizchevroil negotiations were beset by many problems typical of other business deals in the former Soviet Union. In particular, proceedings were complicated by an extraordinarily rapid turnover of Kazakhstani oil officials. Chevron has spent over $700m so far on the venture, but decided to reduce the rate of its investment in 1995 mainly because of problems with export.[1] Russia controls the only pipeline outlet for Kazakhstani oil, and has used this control both for political influence with Kazakhstan and to exert pressure on Chevron to join the CPC on terms unfavourable to the US company.[2] The CPC, rather than spreading risk among its other members, wanted Chevron to put up most of the capital for its project and absorb virtually all of the risk in return for a mere 25% share.[3]

Chevron's problems with export grew shortly after the venture started, when Russia began to complain about the presence of mercaptans – unpleasant-smelling sulphur compounds found in Kazakh oil. Although under the Soviets, Russian industry officials had not seemed to object to the same mercaptans, Russian officials claimed they had to limit Tengizchevroil's access to the pipeline until the situation improved.[4] On 17 March 1993, the Russian government signed an agreement with Chevron and Kazakhstan to allow Tengizchevroil to export 65,000 barrels per day (b/d), increasing to 130,000b/d when the joint venture began to address the mercaptan issue.[5] Chevron duly invested $100m in technology to remove mercaptans from the oil, and the de-mercaptanisation equipment has been operational since the end of 1994.[6] Up to early 1996, however, Russia has not even allowed Tengizchevroil consistently to export the 65,000b/d of the initial agreement.[7]

Russian officials claimed that they had insufficient capacity for their own exports, and that foreign exports, such as those from Kazakhstan, would need to be cut in spite of the March 1993 agreement. These claims do not appear to be consistent with earlier Soviet statements that put the capacity of the pipeline from Aktaru to Samara at 10m tons – well above what the Kazakh and Russian exports combined would require. Thus, although Tengizchevroil is capable of exporting much more, it is limited to 40–65,000b/d.

Chevron is presently looking at short-term export alternatives, or swap arrangements, to absorb excess production caused by Russian pipeline limitations.[8] Chevron Chairman Kenneth Derr in an interview in September 1994 estimated that about 1bn barrels could be exported over the life of the project through swap deals.[9] As Tengizchevroil production increases, it must look for longer-term solutions for its export problems as the present system will not accommodate the project's anticipated export needs.

In mid-April 1996, the US oil company Mobil announced that it would purchase a 25% share of Tengizchevroil. According to Mobil, the 25% share came from Kazakhstan's 50% share of the project.[10] Mobil's purchase is probably tied to a positive resolution of the export issue as most analysts doubt that Mobil would become involved in Tengizchevroil without guarantees for export and profit. At the same time, the President of Russia's Lukoil company, Vagit Alekperov, announced that it had also reached preliminary agreement for the purchase of part of Kazakhstan's stake in Tengizchevroil. According to industry analysts, Kazakhstan may need to give Russia a share of the project as part of a package of concessions that will move forward the much-needed export pipeline projects.

The Azerbaijani Consortium
In 1991, a number of foreign companies – including Amoco, British Petroleum, McDermott, Pennzoil, Ramco, Unocal, TPAO and Statoil – began negotiating with Azerbaijan to develop the Azeri, Chirag and Guneshli fields in the Azeri sector of the Caspian Sea. After protracted and difficult talks, the companies were prepared to sign an agreement with Azerbaijan's pro-Turkish, anti-Russian President Elchibey in June 1993 whereby Azerbaijan's State Oil Company (SOCAR) retained a 30% share of the consortium project.[11] When Elchibey's government was subsequently overthrown, the companies found themselves back at the negotiating table with a new government headed by former Communist Party boss Gaidar Aliyev.

Renewed negotiations were painful, laborious and confusing. Oil companies dealt with a succession of government negotiators, and accusations of demands for bribes and high official corruption were common.[12] However, emerging at the end of a long, troubled process in October 1993, the companies signed a new development protocol which excluded the developed part of the Guneshli field. Fields remaining in the reformed project have combined recoverable reserves estimated at 3.3–3.7bn barrels of oil lying some 193km offshore in about 122 metres of water.[13] In the course of subsequent negotiations, in April 1994 SOCAR officially informed its partner consortium companies that it had assigned the Russian oil company Lukoil a 10% stake from SOCAR's own share.

After several more months of complex bargaining, the final project deal – worth an estimated $8bn in capital investments – was signed on 20 September 1994. The signing ceremony evoked a mixed reaction from Russia, with Stanislav Pugach, Chief of the Main Department of the Russian Ministry of Fuels and Energy, attending the ceremony, while the Ministry of Foreign Affairs spokesperson publicly condemned the deal as illegitimate since the legal status of the Caspian Sea had not yet been determined. At one point, then

Russian Foreign Minister Kozyrev allegedly proposed economic blockades against Azerbaijan to punish Baku for signing the deal.[14] In the course of 1994, Moscow also sent a diplomatic note to the UK essentially claiming veto rights over any deal in the Caspian, and warning the UK not to go ahead with the deal.[15] The companies in the consortium are concerned that Russia could easily apply even more pressure through its control over the Volga River, the only way to bring the heavy equipment necessary for the project into the Caspian.

After the signing, the project required ratification by the Azerbaijani parliament. Although on 15 November 1994, parliament passed a resolution on the contract, lawyers reviewing the text of the proceedings found that the resolution did not meet the legal requirements for ratification and that, therefore, the agreement could not be implemented. On 12 December, the Azerbaijani parliament passed the proper legislation and the deal was finalised. Shares in the consortium, according to the agreement at the time of the signing, were as follows:[16]

- SOCAR (Azerbaijan): 20%
- British Petroleum (UK): 17.13%
- Amoco (US): 17.01%
- Lukoil (Russia): 10%
- Pennzoil (US): 9.82%
- Unocal (US): 9.52%
- Statoil (Norway): 8.56%
- McDermott (US): 2.45%
- Ramco (Scotland): 2.08%;
- Delta-Nimir (Saudi Arabia): 1.68%
- TPAO (Turkey): 1.75%.

Given this intricate joint-ownership structure, an industry official described the project as one of the most nightmarish deals he had ever worked on because of the complex decision-making arrangements, and the natural competition among oil companies both within the deal and outside it.[17] Because Azerbaijan was unable to finance its 20% share, SOCAR later transferred another 5% (in addition to the 10% to Lukoil) to TPAO. In spring 1995, after intense competition between Exxon, Iran, Mobil, France's Elf-Aquitaine, Italy's Agip, and Shell, Azerbaijan awarded a 5% share to Exxon.[18] Iran, in particular, had been pushing heavily for the stake, and had exerted political pressure and offered economic incentives. In the end, the US companies vetoed Iranian participation. This was considered a victory for the US government in its efforts to contain Iran and its influence in the former Soviet Union. At that time, US companies held a 43.8% share of the consortium and, as a bloc, are very influential in the group's decisions. At the same time, of course, these companies remain competitors with differing interests, and do not necessarily act together as a national group.

In December 1994, the consortium established the Azerbaijan International Operating Company (AIOC) to oversee the project's day-to-day operation. It will use $8bn in capital investments to produce approximately 3.7bn barrels of oil over the next 30 years. About 70% of the profits from the deal will go to Azerbaijan, which will receive about $300m in bonus payments.[19] One of the greatest challenges for the consortium will be finding a way to export the oil now that the laborious negotiating phase has been concluded. At this point, none of the options for long-term export appear very stable or attractive. They include mainly pipelines that would pass through Iran, through Azerbaijan's arch-enemy Armenia, through politically unstable Georgia or through Russia, giving it a monopoly over most oil exports from the region.

The consortium is just beginning preparatory work for the first 30-month phase of the project, which is expected to cost about $150m and includes exploration, seismic, environmental and engineering studies, and upgrading an existing production platform; production will not begin until late 1996. If implementation proceeds smoothly, production could reach 200,000b/d by 1999; full production is not expected until 2004 at the earliest.[20]

Given the problems involved in negotiating the contract, the companies are proceeding with caution. They have structured the agreement to allow for a way out if any one of a number of problems impedes implementation of the agreement.[21] These potential problems include:

- the spread of the Nagorno-Karabakh war;
- a sharp deterioration in Russian–Azerbaijani relations;
- internal Azerbaijani instability;
- the Azerbaijani government's failure to find a stable export route;
- the Caspian legal regime dispute.

41

Azerbaijan's Karabakh Field in the Caspian Sea

On 10 November 1995, the Azerbaijani government signed a 25-year $1.7bn contract with an international consortium of Russia's Lukoil, Italy's Agip SpA, and US oil company Pennzoil. Lukoil gained a 32.5% stake, with Agip and Pennzoil holding 30% each. SOCAR, the Azerbaijani oil company, took 7.5%.[22] The subject of the deal – the Karabakh field (unrelated to the disputed area of Nagorno-Karabakh) – is about 120km off the Baku coast, and is estimated to contain 85–180m metric tons (or 450m–1.26bn barrels) of crude oil and gas. Production is scheduled to begin in 1997. Some analysts believe that the conclusion of the deal giving Russia the largest share was motivated by Azerbaijan's desire to placate Moscow on both the Caspian demarcation issue, and on Baku's decision to take short-term exports both north and west.[23]

Kazakhstan's Karachaganak Field

Kazakhstan's Karachaganak field is a large oil- and gas field estimated to be about two-thirds the size of Tengiz. In June 1992, Kazakhstan concluded an agreement with British Gas and Agip to restore the field which had deteriorated because Soviet technology had been insufficient to develop the field further. Attempts by British Gas and Agip to conclude a longer-term deal, however, have been plagued by a variety of technical, political and economic complications that typically block or slow down agreements in this region.

On 10 February 1995, Kazakhstan was forced to bring the Russian natural gas enterprise, Gazprom, into the Karachaganak deal. According to industry sources, Russia had threatened to block export of Karachaganak's oil and gas, and to purchase the field's output at no more than 15% of world prices unless Gazprom was included in the project.[24] On 2 March 1995, British Gas, Agip, Gazprom and the Kazakhstanis signed a 'Production Sharing Principles Agreement'. Industry sources suggest that the signing ceremony was delayed because Russian Gazprom officials wanted more compensation and a better deal.[25] As in other cases, Russian officials argued that Soviet technology had initially developed the field and Russia was therefore entitled to a larger share of the profits.[26]

Kazakhstan's Offshore Deposits Development

Kazakhstan's offshore deposits, estimated at about 25bn barrels (similar to Nigeria) and perhaps containing even more oil than the Tengiz field, are another potentially rich source of oil in the region.[27] With investments of around $400–500m, the Caspishelf Consortium

(CC) deal – not to be confused with the CPC – was ratified in December 1993 to explore this 35,000km^2 territory. Although 35 companies originally expressed interest, in the end the CC agreement included Mobil, Shell, Total, Agip, British Gas, British Petroleum and Statoil. The CC's exploratory efforts, which began at the end of 1994, are expected to continue until the late 1990s. The project also includes a feasibility study on environmental protection. Each partner has the option to acquire development rights for two blocks in the area.[28]

Still in the initial stages of implementation, the CC has not experienced as many of the same difficulties that presently complicate other large oil projects, although it has certainly had its share of bureaucratic entanglements, miscommunication, and problems with the Caspian dispute. If the participating companies do discover the amount of oil they expect in the field, the CC is bound to run into the troubles that have dogged the progress of other projects.

IV. EXPORT OPTIONS

Getting Caspian oil out of the region and into the international markets is a key issue that will ultimately deeply affect the political and economic fate of the countries concerned. This is the area in which domestic complications in each of these countries combine both with the strategic competition among the region's powerful neighbours, and with over-arching technical and infrastructural problems, to produce a seemingly intractable puzzle. The substantial profits and power at stake, however, continue to attract a spectrum of investors, entrepreneurs and political players.

Most of the oil-producing countries of the Caucasus and Central Asia plan at least to double their oil production during the next 5–10 years. The condition, capacity and configuration of the existing Russian-controlled pipelines out of the region are inadequate for the significant increase in oil volumes being generated by the many projects begun after the dissolution of the Soviet Union. However, a monopoly by any one country on all future pipelines would give it leverage that neither the Caspian countries nor the international oil companies want to see. The countries fear the political uses of such leverage, and the oil companies fear that competition among export routes would minimise tariffs and provide alternatives in case of political instability or other disruption.

Both Kazakhstan and Azerbaijan are attempting to establish export routes for their new oil projects. The most advanced of the largest projects, the $20bn Tengizchevroil joint venture, is presently exporting 40–65,000b/d through the Russian system. Although the project is capable of exporting over 90,000b/d, Moscow has limited the amount allotted to Tengizchevroil for reasons of political and commercial leverage. Thus, the project is looking for both short-term routes to absorb its present excess production (30–50,000b/d), and a long-term route for 700,000b/d that, unlike the CPC deal in late 1995, is acceptable to both Chevron and Kazakhstan.

According to the terms of the Azerbaijan consortium agreement, which came into force on 12 December 1994, the companies have agreed to export early oil, beginning at about 40,000b/d and projected to increase to 80–100,000b/d in two years. The real key to the success of the project, however, is how to export the medium- and long-term production, and whether to work with Kazakhstani projects to use jointly a single export pipeline. Exactly where such a pipeline will go, who will finance it and who will control it are the key issues on which billions of dollars and the political future of the region depend.

Small- to Medium-Volume Exports
Small- to medium-volume is defined as exports up to 200,000b/d, and therefore includes Tengizchevroil's extra 30–50,000b/d and the initial oil flow from the Azerbaijani project, which the consortium hopes will come on line by the end of 1996, reaching 80–100,000b/d by the end of 1997. The export options for this kind of volume include:

- upgrading existing Russian pipeline segments to carry oil to the port at Novorossiysk, including reversing the line to Grozny;
- building or using an existing, relatively low-capacity pipeline through Georgia to Supsa, a Black Sea port near Batumi;
- upgrading the line from Aktaru to Samara;
- upgrading a segment of the Russian pipeline from Russia to Central Europe;
- swaps with Russia;
- swaps with Iran by barging or tankering oil through the Caspian to connect with a pipeline spur (to be built) to northern Iran, with oil of equivalent value to be shipped out from Kharg Island in the Persian Gulf;
- swaps with Iran (without building the spur) by trucking or sending by rail the oil to northern Iranian refineries.

Those involved in the Kazakhstani and Azerbaijani projects have been considering a combination of these options. The outbreak of the Chechen war and the subsequent destruction of oil facilities in that area foreclosed a possible swap option with Russia for the Azerbaijani consortium in late 1995, and delayed the implementation of a planned upgrade to the existing line from the Russian–Azerbaijani border through Tikhoretsk to Novorossiysk.[1] In theory, Tengizchevroil could also swap oil with Russian refineries near the Kazakhstani–Russian border, but bureaucratic problems and Russian interest in pushing Chevron into the CPC have prevented Tengizchevroil from making successful long-term arrangements of this kind.

After many months of study and discussion, on 9 October 1995 the AIOC announced its decision to export its early oil both north through Russia to the Black Sea port of Novorossiysk, and west through Georgia to the Black Sea port of Supsa. In this way, the participant oil companies and countries hoped to hedge their bets by having two export outlets. Contrary to press reports, the AIOC has not made a decision on long-term export.

The AIOC hoped that a series of formal agreements with Russia could be concluded by 1 November 1995, but, as seems to be standard for these projects, last-minute changes and demands have delayed the signing. Russia has reportedly brought its full bureaucratic weight to bear, involving 11 ministries in the negotiations, and adding several new demands to the original draft agreement.[2] On 18 January 1996, Russian Prime Minister Chernomyrdin and Azerbaijani President Aliyev signed an inter-governmental agreement on early oil. However, a section of Article 10 stipulates that the agreement becomes effective from the date of the last written notification of the parties' fulfilment of 'internal state procedures'.[3] No one really knows what this means, but it may include Duma ratification, or satisfaction of Russian concerns on the Caspian legal regime dispute.

If it ever gets off the ground, the AIOC project will take about a year to complete, and it will cost approximately $50m to reverse the existing line to Grozny and put in the necessary pumping stations. Under the proposed agreement, the project will export up to 5m tons or 100,000b/d. The project to construct north will face substantial bureaucratic and security problems. On the one hand, Russia insists that the consortium give the northern route priority for an agreed amount before moving oil through the western route; and on the other, there is potential for domestic sabotage as a result of the Chechen conflict.

While wrangling with the Russians over the northern route, the AIOC was also negotiating with the Georgians over the western route. This is expected to take 12–18 months to complete, and will have a capacity of 120–150,000b/d. On 8 March 1996, Georgia and Azerbaijan signed an inter-governmental agreement for this route. Participating companies signed related contracts the same day. According to the agreement, AIOC will use existing Azerbaijani and Georgian pipelines, although the consortium will need to upgrade these and build additional lines. Consortium members still have some way to go to overcome Russian sensitivities, other companies' claims of 'exclusive' negotiation rights, bureaucratic confusion, protection arrangements for the new line, and the question of whether to use existing structures or build new ones.

The Kazakhstan and Azerbaijan governments have at one time or another seriously considered short- and medium-term arrangements with Iran to diversify their exports so that they would not be dependent on Russia.[4] Such options, however, involve considerable difficulties. Some industry analysts believe that if a pipeline spur were built in Iran, it could serve as a long-term alternative that would eventually link pipelines from the Caspian to the Persian Gulf. Any arrangements with Iran, however, would be opposed by the US and would put US companies involved at odds with their government's unilateral trade embargo on Iran. The US has rigorously imposed this embargo on Iran, and has worked aggressively to block Iran's access to international finance. At the end of 1995, further sanctions were ratified against foreign entities investing in Iran's petroleum resources. Similar legislation is pending in the House of Representatives and in early 1996, the Clinton administration and Congress were discussing further actions against Iran.

The Long-Term Pipeline Puzzle

Decisions on long-term oil export from the Caspian region will have major strategic, political and economic consequences. If the oil industry's estimates for exports are accurate, the short-term options outlined above will provide sufficient export capacity until about 2000. Thereafter, new solutions will need to be found. An additional factor affecting oil export is the impact of economic growth in the Black Sea and former Soviet area and future changes in the patterns of crude oil and oil product demand.[5] According to Laurent Ruseckas, an Associate of Cambridge Energy Research Associates, economic growth, particularly in Bulgaria, Romania, Ukraine and Turkey, could triple the size of the current market during 1995–2000.[6] Depending on economic growth patterns, major long-term decisions may have to be made as early as 1997 to leave enough time to finance and build the necessary infrastructure for the new route or routes.[7]

Currently, at least ten different long-term routes out of the region are under consideration by the countries and oil companies involved in the Caspian projects. Undoubtedly, many more proposals will be considered as more oil from new projects comes on line. All the options are complicated, and none is trouble-free because they all either pass through politically unstable areas, involve high costs because of distance and terrain, or are politically risky because they offend the strategic sensibilities of one or another of the regional powers. The choice of routes will be subject to myriad combinations of local and regional interests and technical and infrastructural practicalities.

Although theoretically a single pipeline for Azerbaijan and Kazakhstan might be possible, the complexities of the Kazakhstani projects combined with those of the Azerbaijan consortium raise enormous political difficulties. Participating companies and countries have not ruled out the single pipeline option. However, if the volume of oil justifies the cost, multiple pipelines are more desirable – principally because this should increase competition, keeping tariffs lower, while strengthening the independence and sovereignty of the Caspian countries.

One export route presently under consideration passes initially through Russia or the Caucasus (Georgia or Armenia), and then through the Black Sea and/or Turkey to the Mediterranean. Another route would go through war-torn Afghanistan to Pakistan, depending on how the political situation in the region develops. Other routes under discussion, such as through Iran to the Persian Gulf, or through Kazakhstan to China, do not seem to have gained much support from oil companies and countries involved in the largest projects although, given the fluctuating circumstances surrounding oil in the region, this could change. For example, if exploration in China's Tarim Basin is successful, and the great expectations of Amoco and other companies involved in the project are fulfilled, the China pipeline option could be more favourably considered. The Iranian route to the Persian Gulf would probably only become more attractive if there were a dramatic change in Iranian behaviour and political and economic policies.

The routes that traverse the Black Sea are affected by the Turkish Straits issue. For largely environmental and safety reasons, Turkey has been reviewing limits on tanker traffic through the Bosphorus Straits, where more than one-third of the oil from the former Soviet Union now passes.[8] Russia had hoped to increase oil exports from the Caspian region through its Black Sea ports and then through the Straits, but Turkish views may make it difficult to increase exports from current levels. Although Turkey has cited genuine environmental dangers to explain its position on use of the Straits, political and economic considerations may also play a role in Turkish thinking, given the attractive option of a pipeline route across Turkey itself.

One of the ways Russia is trying to avoid the Straits problem is by arranging with Bulgaria and Greece to build a pipeline that would take Russian oil from the Bulgarian port of Burgas to the Greek harbour of Alexandroúpolis, thereby avoiding Turkey altogether.[9] As of early 1996, the project remained in discussion, and the parties have signed only a protocol for further negotiations. The EU, heavily encouraged by the Greeks, may become involved in financing a feasibility study, and possibly the pipeline, depending on how the project develops. A potential stumbling block for EU financing is the environmental danger that a pipeline might pose to the delicate eco-system of the Alexandroúpolis delta area, and to the main tourist routes in the region. In specific terms, the four long-term pipeline proposals most frequently discussed are:

- the CPC route from Tengiz through Tikhoretsk to the port at Novorossiysk;
- a route through Georgia and Turkey to the port of Ceyhan on the Mediterranean (with assorted variations);
- a route south, dipping about 48km into Iran then going through Nakhichevan, Armenia and Turkey to Ceyhan;
- a route through Afghanistan and Pakistan to the Indian Ocean.

The Caspian Pipeline Consortium Route
The Caspian Pipeline Consortium was formed by Russia, Kaza-khstan and Oman in 1992 to build a pipeline to transport oil from the Caspian region to the Black Sea. The CPC's original plan was to take advantage of existing infrastructure by finishing a 1,600km pipeline that had been 60% completed during the Soviet period (but abandoned because of lack of financing). This would link both the Caucasian and Central Asian fields with expanded port facilities at Novorossiysk, whose improvement CPC would finance, at a total cost of $1.2–1.5bn. The new Novorossiysk facilities would be de-signed to avoid some of the seasonal delays that force the present port to shut down for up to two months a year because of bad weather. The northern arm of the pipeline would extend from Tengiz around the northern curve of the Caspian and then continue straight across to Tikhoretsk and on to Novorossiysk. The southern arm would link up with a pipeline from Baku at the Russia–Azerbaijani border, passing through Grozny to Tikhoretsk. This would involve reversing the line from Baku to Grozny.

From 1992 to late 1994, the CPC tried to persuade Chevron to provide most of the financing for a 25% non-decision-making equity share. Chevron found these terms unattractive, arguing that the deal was structured so that it assumed most of the risk for a dispropor-tionately small amount of equity. That equity comprised class B shares that would give Chevron less of a say in decision-making than the other class A shareholders.[10] The company was expected to assume the risk of a throughput guarantee regardless of whether it shipped the oil. Chevron refused the terms, and in the absence of alternative finance, the deal stagnated.

Chevron did offer a revised financing scheme, which included a more proportionate share of risk-taking and decision-making, but Oman and Russia rejected it.[11] The EBRD attempted to broker a similar deal in place of the CPC structure, but was also rebuffed by Russia and Oman. As mentioned earlier, Russia tried to pressure Chevron through Moscow's control over Tengizchevroil's only export outlet pipeline.[12] Meanwhile, the CPC's position was further complicated by factionalism and disorganisation within the Russian government, making it difficult to discern a clear Russian position.[13]

In late 1994 and early 1995, the CPC restructured the deal, dividing it into two phases. The first involves constructing a southern pipeline arm from north of Baku to Novorossiysk, and expanding the port facility. The second phase involves constructing the northern arm, which is most crucial for Kazakhstan's economic interests. The new deal focused solely on the first phase under which Russia and Kazakhstan agreed to throughput guarantees and to transfer their pipeline assets to CPC, while Oman agreed to raise the $500m finance for the project. The new structure, as agreed on 11 January 1995, appeared to strengthen the CPC and to give Russia and Oman more control than Kazakhstan over pipeline quotas and fees. This version of the CPC also did not attract the necessary financing.

During this period, the Kazakh government was experiencing some disarray with frequent changes of officials responsible for oil issues and sometimes conflicting messages on policy and developments. In early February 1995, for example, Kazakhstan's President Nazarbayev made several confusing assertions in a televised interview, among them that the CPC problems had been solved and that construction had begun. In fact, the CPC was immersed in seemingly intractable difficulties and nowhere near beginning any kind of building work.[14]

In early 1996, with the CPC financing stalled and the badly needed first phase of the project seemingly going nowhere, Russia faced an increasingly serious bottleneck in the pipeline leading towards the port at Novorossiysk. Russian officials revived talks with Chevron representatives over the more equitable plan that Chevron had originally suggested in 1994, while at the same time softening their public stance on the Caspian demarcation issue. If Russia is unable to clear the bottleneck while proceeding with its agreement with the AIOC to export up to 5m tons, it will be forced to refrain from exporting some of its own oil, which could result in significant losses. Alternatively, Russia could again limit Tengizchevroil's exports as it did earlier in the project.

Another reason for lack of movement in restructuring the CPC was Oman's resistance to a further reconfiguration of the deals. In early 1996, however, this changed. For a number of reasons, Dutch financier and Omani Oil Company Chairman, John Deuss – who had been closely involved with the CPC – was, by spring 1996, out of the picture. At this point Oman began to take a more flexible line on the restructuring proposal. Although Oman's first proposal in March 1996 was rejected by the Kazakhs and Russians, the original CPC signatories agreed to form a transition committee to discuss how to achieve an acceptable new structure.

On 27 April, Russian President Boris Yeltsin and Kazakhstani President Nursultan Nazarbayev watched as Russia's Minister of Fuel and Energy Yuri Shafranik, Kazakhstan's Minister of Oil and Gas Nurlan Balgimbayev, and representatives of the Russian state pipeline company Transneft and of eight other oil companies signed a protocol to restructure the CPC with the following allocation of shares:[15]

- Russia: 24%
- Kazakhstan: 19%
- Oman: 7%
- Chevron (US): 15%
- Lukoil (Russia): 12.5%
- Mobil (US): 7.5%
- Rosneft (Russia): 7.5%
- Agip SpA (Italy): 2%
- British Gas (UK): 2%
- Oryx (US): 1.75%
- Munaigaz (Kazakhstan): 1.75%

Although the signing of the protocol represents a breakthrough, participants face months of complicated tariff and tax negotiations and other unresolved issues. Several participants are also casting a wary eye on this summer's Russian presidential elections, which could have a significant impact on the course of the CPC, regardless of what deals are struck before then. Russia and Russian companies now have a 44% share of the deal, but it is difficult to predict how a new Russian government will view the CPC arrangements.

If the CPC deal were successfully concluded, it would probably take $1.2–1.5bn and at least three years to arrange the financing and build the pipeline for the first phase, upgrade and expand port facilities at or near Novorossiysk, and set up the pipeline arm from north of Baku to the Novorossiysk facilities. When completed, though, this would provide Russia with the desperately needed expansion of the Tikhoretsk–Novorossiysk line, thereby relieving a critical bottleneck for Russia. The pipeline would also serve many other projects in the region. According to the negotiators, CPC participants in the project have already booked virtually 100% of the pipeline's capacity.[16] There is some question, however, of whether the port at Novorossiysk and the Bosphorus can accommodate all of this oil, and some industry analysts have stressed that it will be necessary to build an additional line.

From Baku via Georgia and Turkey to the Mediterranean
Georgia offers several options for the long term. First, a possible two-stage project could begin with the short- to medium-term phase mentioned above from Azerbaijan to the Black Sea, followed by a larger-capacity line to carry oil from the Caspian to Tbilisi, and then across Turkey to the Mediterranean port of Ceyhan. Although supporters of this option would also like to link it with the Tengiz project by building a pipeline along the floor of the Caspian to Baku, this seems doubtful given the Caspian legal regime dispute.

Members of the Azerbaijani consortium have been discussing the deal with Georgian President Shevardnadze and his officials. Georgia is keenly interested in the long-term pipeline, seeing it as a way to alleviate its economic difficulties, facilitate economic reform and stabilise the country. But like most other oil projects in the region, the Georgian option has had its share of confusion. In early 1995, no less than three companies were claiming 'exclusive rights' to arrange a pipeline deal through Georgia – all three held documents allegedly signed or 'approved' by Georgian officials.[17] The Georgians are trying to resolve the organisational problem by creating a Georgian International Oil Corporation (GIOC), with George Chanturia as President to coordinate the oil effort. They appear to have eliminated middlemen dispensing 'exclusive rights'.[18]

A variation of the Georgian option could include Armenia, if the Nagorno-Karabakh war continues to abate and the sides come to a peace agreement. Proponents of this variation include former Nagorno-Karabakh negotiator John Maresca, who argues that such a pipeline could be used as an incentive for opponents to conclude the war, and to ensure stability and prosperity for the Armenians and Azerbaijanis.[19] Someone must pay for the pipeline, however, and given the deep animosities that persist in the region, continued sporadic fighting, the lack of governmental control over areas the pipeline would traverse, and the difficult topography and engineering involved, potential backers have hesitated.

Not only is the area such a pipeline would cross more mountainous, and therefore more expensive to construct than other options, but Iraq retains throughput rights in the line that was originally envisaged for use as part of the route through Turkey. Although some analysts and officials have suggested that US and Western governments might offer finance for political reasons, the political will may not exist for such an expensive and potentially risky venture at this time.

Turkey is clearly interested in a long-term pipeline traversing its territory, and has been trying to work with all interested parties to make such an option more attractive to the Azerbaijani consortium than alternative routes through Russia. Then Turkish Prime Minister Tansu Ciller emphasised her strong support for the Georgian–Turkish option during a visit to Baku in summer 1995.[20] Turkish officials introduced a plan later the same year to make the tariff structure more financially attractive than the non-Turkish options. Ankara has reportedly offered to guarantee purchase of up to 100,000b/d for its domestic market. While it seems likely that Russia would prefer a pipeline running mainly through its own territory, Moscow has apparently not yet rejected Turkey's offer of a substantial share of the deal in return for cooperation. Another variation of this option, proposed by the US company Brown and Root, is a pipeline that would loop through Russia before passing through Georgia and on to Turkey.[21]

The Turkish option, however, may encourage Kurdish terrorist attempts at sabotage. Terrorists have chosen oil facilities as targets in the past, and would probably be tempted to do so again, given the political and economic importance of the line to the Turkish government. Oil company officials have confirmed that this possibility is one of many factors they will consider when making pipeline decisions.[22] Such concerns are also bound to be considered by potential financial backers. As indicated earlier, the situation is mitigated somewhat by a decline in Kurdish terrorism and the Turkish government's offer to provide security for the pipeline.

Via Iran and/or Armenia and Turkey to the Mediterranean
A third option frequently discussed is one that would take Azerbaijani oil through a 45km stretch of Iran, with a parallel line passing an equally short distance through Armenia, subsequently combining in a single line on to Turkey. Since both Armenia and Iran are

unstable, it is proposed that one would act as a back-up against disruption of the other. Such a parallel-pipeline scheme would also reduce the leverage of both Iran and Armenia who might otherwise use territorial control over the export line for political purposes. The advantages of the dual route include: improved security by virtue of twin options through a difficult area; increased stability of an Armenian–Azerbaijani peace agreement; the provision of an alternative to a Russian route; and encouragement to Iran for cooperating with Azerbaijan on the Caspian demarcation issue.

The major disadvantage lies in allowing Iran to participate in a lucrative pipeline scheme that would give Tehran some political leverage. Additional questions surrounding this option concern financing. Given the difficulties of arranging such an option and the possible effect pending US legislation with respect to Iran might have on the entire pipeline, oil companies and countries involved did not seem to be pursuing this option actively at the beginning of 1996.

Through Afghanistan to Pakistan
The least likely route would run from Turkmenistan through Afghanistan to Pakistan. For a variety of political, economic and commercial reasons, this route will probably not be feasible until well into the twenty-first century, although it has been a subject of active discussion. Those looking into this option hope that an Afghanistan–Pakistan pipeline could be connected to lines coming from Azerbaijan and Kazakhstan. On 21 October 1995, Unocal Corporation and Saudi Arabia's Delta Oil Company signed an agreement with Turkmenistan that could one day provide an export terminal at the Indian Ocean for Turkmen oil – and conceivably for oil from the rest of the region. The agreement was for a one-year $10m study on an $18bn project to transport gas and oil nearly 1,300km from Turkmenistan to Pakistan. Some analysts believe that even though the 17-year Afghan War continues to rage, parts of Afghanistan controlled by certain warlords might be stable enough to accommodate a pipeline. Although the plan centres on supplying Turkmenistan's substantial gas reserves to a high-demand energy market in Pakistan, the oil pipeline also figures prominently in the design. Although the project is a long shot and is only at the research stage, Unocal and Delta are at least making a small investment in the long-term future. Both companies recognise that financing the project will not be easy under present conditions, but they are hoping that one day it may be possible and that they will be there to cash in on the opportunities.[23]

V. PROSPECTS FOR THE REGION

The future of the Caucasus and Central Asia depends substantially on any changes in the wider geopolitical context in which it is set. Of particular relevance will be the direction of Russia's reform, Tehran's future role in international terrorism and its pursuit of weapons of mass destruction, whether Iran remains unattractive to international investment, how Turkey deals with its separatist problem, and whether the Afghan domestic situation improves. In addition to these external factors, the viability of regional oil development depends also on the operational stability of the Caspian state governments themselves. Although these countries seem relatively settled in early 1996, they are structurally fragile – notably in their dependence on factors beyond domestic control. Both Azerbaijan's President Aliyev and Georgia's President Shevardnadze, for example, appear to have emerged successfully from internal political struggles, but they still have strong and persistent enemies.

Russia
In 1996, most countries involved in Caspian oil development are watching Russia nervously, particularly its evolving domestic political situation. The June presidential elections could have a major impact on Russia's relations with the Caspian countries. Russia's immediate neighbours continue to worry about Moscow's more assertive policy towards the rest of the former Soviet Union, and its hitherto considerable sensitivity to the presence of other actors in what Russia considers its backyard. Nearly all the Caspian countries have expressed concern about the current Russian regime's push for further integration into the Commonwealth of Independent States (CIS), which many observers see as an attempt by Moscow to reassert Russia's role in the region.[1] Yeltsin's decree on 14 September 1995 on 'The Establishment of the Strategic Course of the Russian Federation with Member States of the CIS' has intensified these concerns in various parts of the former Soviet Union.

Whatever the configuration of the Russian government after the 1996 elections, Moscow will continue to be pulled in several directions on the oil issue. Some Russian officials, including First Deputy Prime Minister Oleg Soskovets and Fuel and Energy Minister Shafranik, have expressed the view that Caspian countries owe Russia a debt for its role in developing the region's oil resources under the Soviets, and clearly believe that Russia has a strategic imperative to control such resources.[2] These and other officials have suggested that Russia may wish to assert its regional influence in a number of ways including: closer control over transit on the River Volga, which is crucial for transporting the heavy oil equipment necessary for further development; control of the only existing oil pipeline out of the region; creating legal delays by invoking the Caspian legal regime dispute; and strengthening Russia's political and military influence in the Caucasus.[3]

However, recognition among even the most conservative Russians of the political, economic and technological value of their country's commercial relationship with the West may temper Moscow's future stance in the region. Russia relies on joint ventures with Western partners to develop its energy resources. Its present levels of technology are neither capable of the kind of oil development currently envisaged in the Caspian nor, indeed, in Russia itself. Even those most dismayed by the forfeiture of Soviet control over Union energy wealth are divided about how to regain influence. Russia's Chechnya experience has shown that a heavy hand can be counter-productive and undermine long-term plans. Furthermore, some high-level Russian officials may have personal financial interests in the success of oil development and export projects.[4] Clearly, predicting Moscow's behaviour towards the Caspian is a complicated enterprise, and the emerging picture will no doubt be varying shades of grey rather than black and white. The key for a country interested in the independence and sovereignty of the Caspian states will be to tailor its policies to encourage Russia to allow these countries to develop their energy reserves freely.

Iran
Most of the Caspian countries' decisions on oil policy involve complex trade-offs between conflicting objectives. For example, if Russian behaviour towards these countries becomes more aggressive, and unrest in the Caucasus and Afghanistan remains problematic, an export route through Iran could become more attractive. Iran has assiduously courted countries and oil companies involved in the region, and has offered a range of incentives for involvement in various oil projects. Exchanging dependence on Russia for dependence on Iran – arguably a worse prospect for regional leaders – would be a difficult dilemma for the countries concerned.[5]

For the US, the Russia–Iran dilemma is a conflict between the desire to contain Iran and to foster the independence and sovereignty of the Caspian states. Equally, the US position depends on Iran's future support for terrorism and its efforts to develop weapons of mass destruction. With no easy answers, policy-makers must tread carefully to avoid outcomes that could jeopardise long-term interests in peace, stability, independence and democratic and economic reform. If Russia took a hard territorial line, and the western pipeline option were foreclosed because of war in Georgia or Nagorno-Karabakh, the Caspian countries might be able to send a limited amount of oil out through Iran without compromising their independence. However, doing so could have an impact on their relations with the US. The pipeline option of short parallel sections running through both Armenia and Iran might be viable, but everything would depend on the specific details.

China
In the short term, China will probably not become as deeply involved as other countries in Caspian oil development and export. China's oil import needs, however, may increase its interest in the region's pipeline projects. China will continue to increase its trade and economic ties with Caspian countries, particularly those in Central Asia, and may become the primary trade partner for Kyrgyzstan and Kazakhstan. For their part, the Central Asian countries will base deeper involvement with China both on their own economic and political needs, and on China's foreign-policy behaviour toward its other neighbours. Perhaps the most interesting Chinese–Caspian joint oil venture depends on whether the Tarim Basin holds significant reserves and if so, how that oil could be exported. The Basin's potential, however, will probably not be more fully investigated for at least a decade.

Turkey
Turkey is clearly an influential regional player in the Caspian. The evolution of events there will significantly affect the future of the region's oil developments – particularly the route of a long-term oil pipeline. Much will now depend on Turkish Prime Minister Mesut Yilmaz's handling of a very complex domestic situation. Tensions between government and Kurdish separatists persist, and the government's approach to the Kurdish issue will be a major factor in the long-term pipeline decision, given the potential dangers posed by Kurdish terrorists. Kurdish terrorism has, however, declined over the past two years, and the Turkish government has promised to secure a pipeline through its territory. As part of its support for multiple export options in the Caspian region, the US government

has endorsed the Turkish route as one of several acceptable to the US for oil export from the Caucasus and Central Asia. Other Western governments, on the other hand, have taken the view that export routes are a commercial issue, and have not articulated a public preference on possible pipeline developments.

Afghanistan and Pakistan

At the end of 1995, Afghanistan's 17-year civil war entered yet another stage when *Taleban*, a student Muslim group that emerged in 1994, tried to take control of the capital, Kabul. *Taleban* rebels captured the city of Herat in western Afghanistan in early September 1995, and are pushing to control more territory. The Pakistani government is believed to support *Taleban* for strategic, ethnic and political reasons, while India and especially Iran are believed to be aiding the beleaguered Afghan President Burhanuddin Rabbani in the struggle for the Afghan capital.[6] India would not like to see its historical rival, Pakistan, emerge victorious with the *Taleban*, or the establishment of a dominant Pakistan-backed group in Afghanistan. If the *Taleban* rebels are successful, Pakistan may be in a better position to move forward with the oil and gas pipeline scheme it has advocated for several years. On the other hand, a victory built principally on Pakistan's support without internal consensus would probably not be sustainable. Turkmenistan's project with Unocal and the Delta Oil Company is an attempt to sound out the possibilities in a highly volatile situation with an eye to the long-term future. Without an Afghan peace settlement, the project is unlikely to progress much beyond the study stage in the near future.

Pakistan itself is troubled by endemic corruption, large budget and trade deficits, a faltering economy and political instability. Prime Minister Benazir Bhutto is struggling with the difficult issues of the role of Islam and the status of the army in Pakistan's government and society. Pakistan's economic problems will make it difficult for the country to participate in financing large projects such as the pipeline from Turkmenistan, and, in any case, instability inside Pakistan could make attracting outside financing difficult.

CONCLUSION

For two centuries, oil has been the focal point for regional and international competition in the Caucasus and Central Asia. For external and internal players alike the competition has been established over years of intermittent conflict. Culturally embedded issues continue to lead to inter-ethnic fighting, but administrative corruption and under-developed commercial and legal practices also play a role. These will undoubtedly have an impact on future ventures to develop Caspian oil.

Ultimately, the development of the region's oil reserves depends on a flow of financial and technological support from Western partners, as well as the emergence in the newly independent regional states of stable political environments and standardised legal and commercial infrastructures. These developments, in turn, depend on establishing stable relations between the regional states and outside interests.

Russia has a profound impact on this process. Moscow continues to exercise significant influence in the region, despite internal disagreements over Russia's strategic and commercial interests in the Caucasus and Central Asia, which lead to fragmented Russian policy implementation. Even those Russians most unhappy with the loss of Soviet-era control over the energy wealth in the Caspian Basin are divided about how to regain influence. Russia itself relies on joint ventures with Western partners to develop its energy resources as its existing technology for oil exploitation is not capable of the kind of oil development currently envisaged in the Caspian. Moreover, Russia's Chechnya experience shows that bringing military force to bear can be counter-productive, undermining long-term plans. The alleged financial interests of high-level Russian officials may also play a role in the success of oil development and export projects. Clearly, predicting Moscow's behaviour towards the Caspian is a complicated enterprise, and the policies of outside actors will have to be tailored to promote more positive Russian behaviour.

Western countries – predominently the US, the UK, France and Turkey – share some common goals in the region. These countries support the independence of the states of the Caucasus and Central Asia, emphasising that oil is the key to their economic viability and sovereignty in the face of Russian efforts to exert heavy influence in the region.

Multiple short- and long-term routes promote benefits for the outside partners, encouraging commercial competition, keeping tariff rates lower, promoting fairer commercial practices, and safe-guarding exports against interruption. For example, US support for a route to Turkey came about not only because it reduces the Caspian countries' dependence on a single route through Russia, but also because it allows exporters to avoid weather and capacity problems at the Russian port of Novorossiysk and reduces the potential for oil spills and tanker accidents in the Black Sea and Turkish Straits. It also reduces the pressure for a route through Iran to the Persian Gulf, a factor especially important to the United States.

Policy Options

With the clear geostrategic goal of opening the Caspian to oil exploitation and export, Western and other countries must cope successfully with the uncertainties of Russian policy and regional economic and political development. The outside partners can take several steps to reinforce the efforts of the Caucasian and Central Asian countries to consolidate their independence and diversify control over their oil resources.

First, and most important, all of the partners, both in and out of the region, should support multiple routes for short-, medium- and long-term oil export, thereby avoiding dependence on any one country for an outlet. This option is difficult because it requires substantial financing, which is a decision usually based on financial risk rather than strategic interest. Nevertheless, Western government-backed guarantees could provide a solution to this problem, if accorded sufficient priority by the partner states. Even if Western partners do not provide financial guarantees, they should continue to offer high-level political and other support – such as infrastructure contributions and technical expertise and economic aid – to promote multiple pipelines.

Second, the Western countries should coordinate their policies on non-competitive issues more closely. Oil issues should figure prominently on the agendas of high-level meetings involving the major players – the US, the UK, France, Italy, Norway, Saudi Arabia, Russia, Turkey, Azerbaijan, Kazakhstan, Georgia, Turkmenistan, Uzbekistan and Oman. Assistance to the energy sector should be more closely coordinated to avoid duplication, with information-sharing to achieve a commonly approved set of regional priorities. A concerted effort by the West to raise concerns, particularly with Russia, about fair commercial practices might be more effective than a set of uncoordinated approaches.

Third, more assistance should be made available to develop the region's energy infrastructure, legal framework and technical expertise – particularly in Georgia and Kazakhstan, but also in Azerbaijan, Uzbekistan and Turkmenistan. The international financial institutions in particular should fine-tune their aid programmes to become more involved in energy projects that will enhance the viability of the Caspian Sea states. Such projects should be seen as positive, offering tangible and visible results.

Fourth, all partners should continue to emphasise principles of common access, transparency and non-discriminatory pricing, all of which have been laid down and agreed in the European Energy Charter. This would be another way to encourage Russia to abide by accepted international standards in dealing with its neighbours in the region, and it would also smooth the path of Western commercial involvement there.

Fifth, the Western countries should support the rights of Azerbaijan, Kazakhstan and Turkmenistan to develop projects in the Caspian Sea, at the same time encouraging proper environmental safeguards. This relates to the Caspian demarcation issue, which Russia has sought to use to strong-arm other Caspian states. Support for the non-Russian Caspian states would further their efforts to consolidate independence.

Sixth, partner states should support Western commercial activities in the region. Western companies need substantial political support to manoeuvre successfully in the Caucasus and Central Asia. It is in the interest of Western government to foster the deep involvement of their domestic firms in the exploitation of this highly strategic resource.

Finally, the United States should lift Section 907 of the Freedom Support Act restricting aid to Azerbaijan. It hampers US ability to influence a country with a vital strategic resource, and also limits US effectiveness in working with Azerbaijan and Armenia in the peace negotiations over Nagorno-Karabakh.

These steps should reinforce the efforts of the countries in the Caucasus and Central Asia to achieve independence while diversifying control over their oil resources. Although the final trajectory of the region's development is difficult to predict, the countries with strategic interests in exploiting its oil resources should not simply respond to the constant shifts in policy of the regional states. Instead, they should remain focused on long-term goals and take advantage of the historic opportunities that the region offers.

Notes

Introduction

[1] Oil reserve estimates for the region vary greatly and range from 30bn to 200bn barrels. These estimates include proven and possible reserves. Industry analysts often use a middle-range figure of 90bn, similar to China or Mexico.

[2] The US and Mexican reserves cited here are proven reserves as opposed to estimated reserves for the Central Asian and Caucasus region.

[3] See Daniel Yergin and Joseph Stanislaw, 'Oil: Reopening the Door', *Foreign Affairs*, vol. 72, no. 4, September/October 1993; and Jay Bhutani and Ed Morse, 'Pushing Against the Limits: What Oil Markets May Look Like by 1998', *Petroleum Intelligence Weekly, Oil Market Outlook*, January 1994.

[4] Eugene Khartukov and Olga V. Vinogradova, 'Former Soviet Union: Another Poor Year', *World Oil*, vol. 215, no. 8, August 1994. Fareed Mohamedi, 'Pipeline Politics in Central Asia', summary of a presentation given at the American Petroleum Institute's Pipeline Conference, Dallas, TX, 25 April 1995. Steve LeVine, 'High Stakes', *Newsweek*, 17 April 1995.

Chapter I

[1] For an excellent account of the history of oil in the region, see Daniel Yergin, *The Prize* (New York: Simon and Schuster, 1991).

[2] *Ibid.*, p. 59.

[3] Robert W. Tolf, *The Russian Rockefellers* (Stanford, CA: Hoover Institution Press, 1976), pp. 50–60.

[4] Nikolai Baibakov, *The Cause of My Life* (Moscow: Progress Publishers, 1984). Baibakov's memoirs discuss his role in destroying the Caucasus oil-fields to deny Germany their use.

[5] Yergin, *The Prize*, p. 337.

[6] *Ibid.*, p. 115.

[7] Tolf, *The Russian Rockefellers*, p. 117. On the present situation, see J. Robinson West, 'Pipelines to Power', *Washington Post*, 8 June 1994.

[8] Yergin, *The Prize*, p. 115. On the present situation, see Steve LeVine, 'Oil Consortium to Skirt Russia in its Shipments', *New York Times*, 7 October 1995; Steven Erlanger, 'A Corrupt Tide in Russia from State-Business Ties', *New York Times*, 3 July 1995; John Thornhill, 'Russia's New Oil Riddle', *Financial Times*, 6 June 1995; Paul Klebnikov and Dana Wechsler Linden, 'So Sue Me', *Forbes*, 1 August 1994.

[9] Tolf, *The Russian Rockefellers*, pp. 156–60.

[10] *Ibid.*

[11] Khartukov and Vinogradova, 'Former Soviet Union'.

[12] See Robert O'Connor, Richard Castle and David Nelson, 'Future Oil and Gas Potential in the Southern Caspian Basin', *Oil and Gas Journal*, 3 May 1993, pp. 117–26; Michael J. Strauss, 'Caspian Sea May Offer Wealth of Oil and Gas, Geologists Say', *Journal of Commerce*, 16 September 1991, p. 6B; Khartukov and Vinogradova, 'Former Soviet Union', p. 69.

[13] 'Kazakhs Studying Revised Terms for '95 Bidding Round', *Platt's Oilgram News*, 5 December 1994, p. 1.

[14] 'Turkmenistan Issues Ambitious Oil, Gas Programs', *Oil and Gas Journal*, 8 November 1993, p. 100.

[15] Khartukov and Vinogradova, 'Former Soviet Union'.

[16] *Ibid.*, p. 69.

Chapter II

[1] 'A Hero of Our Time', *The Economist*, 12 February 1994, pp. 55–56; Misha Glenny, 'The Bear in the Caucasus: From Georgian Chaos, Russian Order', *Harper's Magazine*, March 1994.
[2] *Ibid.*
[3] S. Rob Sobhani, 'Russia Tests the US in Azerbaijan', *Wall Street Journal*, 28 June 1994.
[4] *Ibid.*; Lally Weymouth, 'Azerbaijan – Who'll Stop the Russians?', *Washington Post*, 11 October 1994; Laura Le Cornu, 'Azerbaijan's September Crisis: an Analysis of the Causes and Implications', Former Soviet South (FSS) Project Briefing no 1, Royal Institute of International Affairs (RIIA), London, January 1995.
[5] Stephen J. Blank, *Energy and Security in Transcaucasia* (Carlisle, PA: US Army War College, 1994); Thomas Goltz, 'The Hidden Russian Hand', *Foreign Policy*, no. 92, Autumn 1993, pp. 92–116; Sobhani, 'Russia Tests the US'.
[6] Mary Bruckner Powers, 'Balkan Intrigue Sets Routes', *ENR*, 23 October 1995, p. 15; Le Cornu, 'Azerbaijan's September Crisis'.
[7] Steve LeVine, 'Moscow Pressures its Neighbors to Share Their Oil, Gas Revenues', *Washington Post*, 18 March 1994.
[8] *Ibid.*
[9] *Ibid.*
[10] LeVine, 'High Stakes'.
[11] Robert V. Barylski, 'Russia, the West, and the Caspian Energy Hub', *Middle East Journal*, vol. 49, no. 2, Spring 1995.
[12] LeVine, 'Moscow Pressures its Neighbors'.
[13] Private communication with US oil company executive, 29 August 1995.
[14] Steve LeVine, 'Kazakhs Accuse Moscow of Stopping Oil Exports', *Financial Times*, 28 June 1994.
[15] See letter to the editor from the Ambassador of Azerbaijan to the United States, Hafiz Pashayev, *Washington Times*, 21 October 1994, in which he refers to the efforts of both President Clinton and Vice-President Gore in pressing the Russians to support Azerbaijan's International Oil Consortium.
[16] See Mike McCurry, White House briefing transcript, 2 October 1995.
[17] The Nagorno-Karabakh war has deep roots in the history of the region and both sides bear responsibility for the many atrocities that have occurred. Although Armenia officially denied involvement in the war for a long time, it has clearly played a key role in the conflict.
[18] While 907 does not overtly ban direct assistance to refugees, it effectively does so since only Azerbaijani government agencies are generally capable of storing and delivering more than small quantities of aid. Most non-governmental aid organisations operating in Azerbaijan are dependent on government facilities.
[19] Gareth Winrow, 'Turkey in Post-Soviet Central Asia', FSS Project, RIIA, London, 1995, p. 44.
[20] See later section on long-term pipeline routes for more detail.
[21] See *Kazakhstanskaya Pravda*, 14 May 1992.
[22] *Nezavisimaya Gazeta*, 21 June 1994, in *Current Digest of the Soviet Press*, vol. 46, no. 25, June 1994, p. 22; Nancy Lubin, 'Islam and Ethnic Identity in Central Asia: A View from Below', in Yaakov Ro'i (ed.), *Muslim Eurasia: Conflicting Legacies* (London: Frank Cass, the Cummings Centre Series,

1995), pp. 66–67.

[23] LeVine, 'High Stakes', p. 13.

[24] Oles M. Smolansky, 'Turkish and Iranian Policies in Central Asia', in Hafeez Malik (ed.) *Central Asia* (New York: St Martin's Press, 1994), pp. 283–84.

[25] Daniel Southerland, 'Azerbaijan Picks Exxon Over Iran for Oil Deal', *Washington Post*, 11 April 1995; also two private interviews with Western oil industry officials, 9 September 1995.

[26] Smolansky, 'Turkish and Iranian Policies', p. 291.

[27] *Ibid.*

[28] Helen Boss, 'Turkmenistan: Far From Customers Who Pay', FSS Briefing no. 4, RIIA, November 1995.

[29] 'Turkmen Energy Council Sets Up; Agreements Signed with Iran, Russia', World Service, Moscow, 5 April 1994, in *BBC Summary of World Broadcasts*, 15 April 1994.

[30] Ahmed Rashid, 'Chinese Challenge', *Far Eastern Economic Review*, 12 May 1994, p. 30.

[31] Ross Munro, 'Central Asia and China', in Michael Mandelbaum (ed.), *Central Asia and the World* (New York: Council on Foreign Relations Press, 1994), p. 232.

[32] *Ibid.*, pp. 230–32.

[33] *Ibid.*, p. 230.

[34] *Ibid.*

[35] Mamdouh G. Salameh, 'China, Oil and the Risk of Regional Conflict', *Survival*, vol. 37, no. 4, Winter 1995–96, p. 133.

[36] *Ibid.*, p. 139.

[37] Steve Coll, 'Central Asia's High-Stakes Oil Game', *Washington Post*, 9 May 1993, pp. A1, A28.

[38] Salameh, 'China, Oil', p. 139.

[39] 'China's Upstream Programs Advance Onshore and Offshore', *Oil and Gas Journal*, 25 September 1995; Paul Post, Alex Milne and Li Bingjian, 'Exploration Potential of Areas in Onshore China, Third Round', *Oil and Gas Journal*, 11 September 1995.

[40] Petroleum Finance Company, 'Pipeline Politics in Central Asia and the Caucasus', *Focus on Current Issues*, May 1993, pp. 21–22.

[41] David Nissman, 'Ethnopolitics and Pipeline Security', *Prism*, 6 October 1995, p. 8.

[42] Christopher Pala, 'Russia's Caspian View Violates Law, in Eyes of Kazakhstan's Minister', *Platt's Oilgram News*, 22 November 1994, p. 1.

[43] Discussions with international legal experts and US oil company officials, 26 June, 12 September and 20 October 1995; preliminary opinion prepared by Western law firms for AIOC, January and April 1994.

[44] Pala, 'Russia's Caspian View'.

[45] Murray Feshbach, *Ecological Disaster: Cleaning Up the Hidden Legacy of the Soviet Regime* (New York: Twentieth Century Fund Press, 1995); interview with Feshbach, 23 January 1996.

[46] 'Niyazov Proposes Caspian Oil Fields Consortium', *Interfax*, 20 October 1994.

[47] *Ibid.*

[48] Le Cornu, 'Azerbaijan's September Crisis'; Arif Useinov, 'Baku Fears Economic Sanctions from Teheran', *Current Digest of the Post-Soviet Press*, vol. 47, 10 May 1995, p. 22.

[49] Interview with Western oil industry expert, 18 August 1995.

[50] Erik Peterson and Anthony Gaita, 'Danger in the Caucasus: Pipeline Politics are Leading Yeltsin into a Potential Quagmire in Chechnya...',

CSIS Watch, January 1995.
[51] See Robert Ebel, 'The History and Politics of Chechen Oil', *Post-Soviet Prospects*, vol. 3, no. 1, January 1995.
[52] LeVine, 'High Stakes'.
[53] Zina Moukheiber, 'Mr. Five Percent', *Forbes*, 4 July 1994, p. 74; Christopher Dickey, Steve LeVine and John Barry, 'Carpetbaggers of Kazakhstan', *Newsweek*, 17 April 1995, pp. 13–14.
[54] *Ibid.*; Steve LeVine, 'US and Russia at Odds Over Caspian Oil', *New York Times*, 4 October 1995.
[55] Louise Hidalgo, 'Western Firms Battle to Be the Oil Barons of Baku', *The Times*, 23 February 1994.
[56] 'To Our Clients and Friends: Kazakhstan Oil and Gas Legislation, Recent Developments', letter from the law firm Debevoise and Plimpton, 5 May 1994.
[57] Personal communication with a Republic leader, July 1991.
[58] Dickey, LeVine and Barry, 'Carpetbaggers'.
[59] Geoff Winestock, 'Banks See Kazakhstan as a Golden Opportunity; Oil Wealth Draws Investors', *Journal of Commerce*, 6 May 1994, p. 1A.

Chapter III
[1] Raymond Bonner, 'Getting This Oil Takes Drilling and Diplomacy', *New York Times*, 15 February 1995.
[2] LeVine, 'Moscow Pressures its Neighbors'; and Petroleum Finance Company, 'Pipeline Politics II', *Focus on Current Issues*, April 1994.
[3] Petroleum Finance Company, 'Pipeline Politics in Central Asia and the Caucasus'.
[4] LeVine, 'Moscow Pressures its Neighbors'.
[5] Petroleum Finance Company, 'Pipeline Politics in Central Asia and the Caucasus', pp. 5–7.
[6] Information provided by Chevron, 18 July 1995.
[7] LeVine, 'Moscow Pressures its Neighbors'.
[8] 'Chevron Might Seek Permission to Deal with Iran', *New York Times*, 12 September 1995.
[9] Robert Corzine, 'Chevron Upbeat on Tengiz Project – The Kazakhstan Scheme Would Virtually Double the Group's Oil Reserves', *Financial Times*, 27 September 1994, p. 27.
[10] Sander Thoenes, 'Mobil to Buy 25% of Big Asian Oilfield', *Financial Times*, 18 April 1996.
[11] Gavan McDonell, 'The Euro-Asian Corridor', *Post-Soviet Business Forum* (London: RIIA, 1995).
[12] 'Lawyer Predicts Investigation of Consortium', *Oil Daily*, 3 November 1995; interviews with Western oil company officials, consultants and energy specialists, spring and summer 1995.
[13] Khartukov and Vinogradova, 'Former Soviet Union', p. 76.
[14] Pala, 'Russia's Caspian View', p. 1.
[15] LeVine, 'Kazakhs Accuse Moscow'.
[16] McDonell, 'Euro-Asian Corridor', p. 40.
[17] Interview with Western oil official, 18 July 1995.
[18] Southerland, 'Azerbaijan Picks Exxon', p. A5.
[19] 'Caspian Accord Gets Approved By Azerbaijan', *Platt's Oilgram News*, 16 November 1994, p. 1; 'Three Fields Slated for Development Off Azerbaijan', *Oil and Gas Journal*, 26 September 1994.
[20] AIOC press release, Baku, Azerbaijan, 9 October 1995.

21 Margaret McQuaile, 'Dispute Aside, Caspian Work Will Start', *Platt's Oilgram News*, 9 November 1994, p. 1.

22 'Azerbaijan Marks More Caspian Sea Progress', *Oil and Gas Journal*, 19 June 1995.

23 Rossen Vassilev, 'The Politics of Caspian Oil', *Prism*, 12 January 1996, pp. 10–12.

24 Steve LeVine, 'Asian Gas Exporters Try To Bypass Russia', *New York Times*, 9 September 1995, p. 1; LeVine, 'Moscow Pressures its Neighbors'; McDonell, 'Euro-Asian Corridor', pp. 38–40.

25 Interview with Western oil company officials, 18 April 1995.

26 Steve LeVine and Robert Corzine, 'Russians to Get Share of Kazakhstan's Largest Gas Field', *Financial Times*, 8 November 1994, p. 3.

27 David Ignatius, 'Pipeline Politics' Odd Alliances', *Washington Post*, 25 June 1995, p. H2.

28 Khartukov and Vinogradova, 'Former Soviet Union'.

Chapter IV
1 Ignatius, 'Pipeline Politics' Odd Alliances'.

2 'Early Oil Exports Find New Hurdle in Changed Language', *Platt's Oilgram News*, 24 January 1996, p. 1.

3 *Ibid.*

4 LeVine, 'High Stakes'; 'Chevron Might Seek Permission to Deal with Iran', *New York Times*, 12 September 1995.

5 Private communication from Howard Chase, Manager External Affairs, Government and Public Affairs, BP Exploration Operating Company, 12 December 1995.

6 'Pipeline Issues Shape Southern FSU Oil, Gas Development', *Oil and Gas Journal*, 22 May 1995.

7 This conclusion is based on interviews from mid-1995 to early 1996 with oil industry consultants, specialists and Western oil company officials.

8 John Pomfret, 'Political Shoals Imperil Role of Turkish Strait', *Washington Post*, 27 April 1995, pp. A29–30.

9 *Ibid.*

10 'Pipeline Politics in Central Asia and the Caucasus', pp. 6–7; speech delivered by Chevron's Richard Matzke at the Kazakhstan International Oil and Gas Exhibit, Almaty, Kazakhstan, October 1994.

11 Interviews with oil industry specialists and Chevron officials in 1995.

12 LeVine, 'Moscow Pressures its Neighbors'.

13 Barylski, 'Russia, The West'.

14 Transcript of 'Adam Smith's Money World', US Television, 16 February 1995.

15 'Caspian Pipeline Deal Gets Closer', *International Herald Tribune*, 29 April, 1996.

16 Sander Thoenes, 'Russia Agrees Kazakh Oil Pipeline', *Financial Times*, 29 April 1996.

17 Interview with George Makharadze, Deputy Chief of Mission of the Georgian Embassy, Washington DC, 24 January 1995.

18 *Ibid.*

19 John J. Maresca, 'Using Oil to Stop a Nasty War', *Journal of Commerce*, 1 August 1994.

20 Remarks by then Prime Minister Ciller to the representative of the Oil Consortium and relevant Azeri authorities, Baku, 11 July 1995.

21 Brown and Root, 'Caspian Sea Oil Export Pipeline', October 1994.

22 Interviews with Western oil company officials throughout 1995 and early 1996.

[23] 'Turkmenistan Draws Western Energy Firms', *International Herald Tribune*, 23 October 1995, p. 15.

Chapter V
[1] Evgeni Novikov, 'Central Asia's Search for Security', *Prism*, 3 November 1995, pp. 8–9; Stanislav Lunev, 'Russia's New Military Doctrine', *Prism*, 1 December 1995, pp. 1, 12; Vladimir Socor, 'This Week in the Regions', *Prism*, 22 September 1995, pp. 1–4.
[2] Barylski, 'Russia, the West'; Patrick Connole, 'What is Russia Up To In The Caspian? UK Finds It Very Difficult to Fathom', *Platt's Oilgram News*, 9 November 1994, p. 1.
[3] *Ibid.*; LeVine, 'Moscow Pressures its Neighbors'; Mohamedi, 'Pipeline Politics in Central Asia'; LeVine, 'High Stakes'.
[4] Peter Fuhrman, 'Robber Baron', *Forbes*, 11 September 1995, pp. 208–20; Erlanger, 'A Corrupt Tide'.
[5] Vassilev, 'The Politics of Caspian Oil', p. 11.
[6] 'Kabul Threatened', *The Economist*, 23 September 1995, p. 27.